PROTECTING YOUR BUSINESS AGAINST *ESPIONAGE*

PROTECTING YOUR BUSINESS AGAINST *ESPIONAGE*

———◆———

Timothy J. Walsh

&

Richard J. Healy

amacom

A Division of American Management Association

PREFACE

Almost everyone enjoys an espionage story. Fictional activities of foreign agents and industrial spies have been highlighted in books, movies, and television or radio programs for many years. Some of the situations and characters portrayed have been humorous, if not ridiculous—for example, the TV series actor who had a transmitter in his shoe. Because of frequent exposure to such portrayals, one might conclude that while stories about industrial espionage are good entertainment, the problem is not encountered in real life. Even if one did believe that information loss is a threat, he might also conclude, in light of the international publicity given such items as the bugged martini olive, that data are lost mainly through electronic eavesdropping devices.

The industrial spy does, in fact, exist—and he does use electronic listening devices. But there are many other ways in which valuable information is lost. All those data loss risks will be emphasized here, but no effort has been made to spice the book with the intrigue, glamour, and sex generally found in cloak-and-dagger thrillers. Instead, the text is intended to be a practical, factual discussion of the hazards faced by businessmen, the types of information to be protected, and the protective techniques that can be used. However, it should be apparent from the material presented that the potential for information loss in real life can be just as hair-raising as it is in fiction.

The reader should understand that every organization, regardless of activity or size, has data that would be of great value to others, and that almost every enterprise is threatened by the specter of information loss. A complacent or incredulous attitude about data loss is an invitation to grave harm to—perhaps even destruction of—the organization. Whether valuable information is stolen by an industrial spy or released inadvertently by a loyal employee who is unaware of the pitfalls in this area, the result may be the same—serious or fatal damage.

Managers in every type of business organization as well as those in public and private institutions must give attention to information protection programs. This book is intended to chart the way by offering a systematic, orderly approach to the problems of design and implementation. In addition to providing practical guidelines for executives in charge of a company security program, the book will serve as a valuable reference for four principal groups.

The first group comprises senior executives, including general managers of divisions, presidents of subsidiaries, and management boards of major enterprises. This group includes all those functional senior executives within whose subordinate organizations specialized confidential and trade secret information is generated.

The second group includes middle-level managers responsible for day-to-day supervision of the work activities of major clusters of employees. This audience must be considered because the conduct of employees and their specific behavior with respect to sensitive or proprietary information will very often determine the degree to which such information is actually protected.

The third group consists of research and development personnel responsible for discovering and perfecting new products and processes. The term "trade secret" will most often apply to the results of this group's activities. Trade secret information, a subset of the whole class of confidential or sensitive information, is particularly relevant for this group and requires special understanding and safeguards to achieve optimum legal and practical security.

The fourth group for whom this book will be useful is made up of general, patent, and house counsel. Most of the existing better-known works on trade secrets are essentially legal in orientation and style. While they allude to techniques for implementing legal concepts, they do not set out much in the way of directly applicable, usable guidelines to adopting those practical techniques. This book seeks to complement the legal texts with intensely practical statements of problems and solutions.

TIMOTHY J. WALSH
RICHARD J. HEALY

CONTENTS

PROTECTING YOUR BUSINESS AGAINST *ESPIONAGE*

1

THEFTS BY
INDUSTRIAL ESPIONAGE

In ancient times, the most valuable possessions of merchants were jewels and precious metals, which were secured in strong, locked chests. Today, the most valuable business assets are no longer that concrete nor, in a severely competitive business environment, is their protection that simple. Secret data have become the most priceless commercial treasures. The constant need for new products and the changes in fashion and the short lifetime of old ones together result in a frantic effort to discover new processes and variations of those already in use. Intellectual information has evolved into an important business asset, and, concurrently, a fundamental problem has developed: the risk of loss through industrial espionage. Protection of intellectual information is an essential element in the management of the present-day enterprise.

One indication of the importance of intellectual information in the United States today is the amount being spent on research and development. It increased from $90 million in 1920 to more than $26 billion in 1971. It follows that an industrial spy may obtain millions of dollars' worth of information at practically no cost. However, the espionage problem is not limited to giant corporations or to organizations with research programs. Even the smallest business is anxious to conceal such information as profit margin, pricing plans, sales volume, future plans, and business problems. The owner or manager may even intentionally mislead his competitors on the state of his business. It has been said that a business that has no information to protect is not competitive.

Protected information can be obtained without employing a professional spy. Alternative ways are discussed in Chapters 4 and 5. Here we may note that the proprietor of a small business such as a supermarket may find that he can bribe one of the employees of a competitor to bring him information about weekly specials or other promotional activities that would give him a competitive advantage. That example illustrates an important point: loss of intellectual information can damage a firm in two ways. First, the expected competitive advantage may be lost and, second, the money and effort spent on research or the collection of data may have been wasted.

It is difficult for the executive whose business practices are ethical to realize that there are unethical individuals who will take advantage of a situation in which intellectual property can be stolen and misused. As a result, many owners and managers tend to close their eyes to the industrial espionage threat and adopt the complacent "it can't happen here" attitude. Business suicide can result.

The word "espionage" is derived from the French verb *espionner,* which means to see or discover something intended to be concealed. As the origin of the word implies, any method may be used. The individual who steals information can be expected to violate all basic personal and property rights in whatever way he finds necessary. On the other hand, the executive who is responsible for any enterprise, large or small, must understand that the threat of loss of valuable information through industrial espionage is real and can, in fact, endanger his own organization.

The hazard may be unappreciated because it is impossible to know the yearly losses due to industrial espionage. In his shadowy, conspiratorial world, the industrial spy must operate quietly and discreetly. A lonely operator, he must not discuss his work with anyone. His greatest success is to make off with the desired information without the victim ever knowing that the theft has occurred. Often the corporate victim that does discover the information loss will assist the spy in concealing the theft through a reluctance to admit to the public or its stockholders that it has been victimized. Nevertheless, a few well-publicized court cases have given the general public a glimpse into the sinister underworld of industrial espionage. Those cases, some of which will be discussed in this book, demonstrate that a real problem exists. Further, they suggest that the magnitude of loss, if known, would be startling. An estimate is that in the United States it may be as much as $4 billion annually.

The problem of protecting intellectual information is complicated by modern communications. We are living in what has been described as an instant society. Mountains of information are made available to most organizations by computers and other means of processing data. Travel between states and countries now requires only hours instead of the days or weeks of only a few years past. Information can be transmitted over great distances in seconds. As a result, stolen information can be transported to a distant state or foreign country in a very short period of time. Moreover, the information can be put to use so quickly that the victim, even if he is aware of the theft, may be unable to take any action that will truly protect his interests.

Although industrial espionage might be thought a modern activity and the industrial spy a twentieth century phenomenon, neither is really new. It is true that industrial espionage came of age with the industrial revolution. As production methods became more complex and the search for new processes intensified, obtaining the trade secrets of competitors became increasingly worthwhile. However, the theft of trade secrets had already been going on for thousands of years. In the ancient cultures also there were processes and techniques that had great value and were carefully protected. One such example is the secret of silk.

THE ANCIENT SILK SECRET THEFT

Silk originated in China at least 5,000 years ago; [1] there is historical evidence that it was an integral part of the culture of the northern provinces in about 3,000 years B.C. For many centuries, the production and processing were maintained as jealously guarded secrets. The exportation of silkworms, eggs, and mulberry tree seeds was prohibited by law, and the penalty for disclosure of the silk secret was decreed to be death by torture. Apparently the decree, along with the fact that silk manufacture was a royal monopoly, was adequate security protection.

The production of silk was under the direct supervision of the royal family until about 1150 B.C. After that time, it was expanded so that silk could be used more generally. Because of the popularity of silk, stories about the remarkable fabric began to filter out to Korea, India, and other neighboring countries. The wily Chinese apparently recognized the danger to their secret that was potential in the release of information and decided that an additional security measure was needed. Accordingly, they created a cover story to be told to foreigners. Their story was that they had merely developed a technique for refining the fleece of sheep. The fleece was first sprinkled with pure water in the sunshine during a particular season and then combed. The result was fine silk thread ready for weaving.

The Chinese security program was apparently very effective, because no foreigners were aware of the true source of silk for about 3,000 years. Eventually two members of the royal family itself made the secret available to foreign countries. The first royal culprit was involved with Khotan, a southern Chinese province near the border of India. In about 140 B.C. the ruling prince of Khotan sent a deputation to the Chinese emperor to ask for some silkworm eggs and some mulberry tree seeds or cuttings. The emperor curtly refused. The prince next sought the hand of the emperor's daughter by proxy. The emperor agreed to the marriage, and the prince of Khotan sent another group of representatives to escort the princess to her new home.

Before he dispatched his representatives to China, the prince of Khotan instructed the leader to have a private conversation with the prospective bride and impress upon her that Khotan had neither silk nor methods of manufacturing it. The usual garments of the women in her future country were made from wool or goat hair. The leader of the delegation was to hint that the princess might want to bring with her the means of producing silk so she could continue to wear the garments to which she had been accustomed.

The princess took the hint and, at the risk of death by torture, concealed mulberry tree seeds and silkworm eggs in her headdress. Legend is that all members of the party were carefully searched when they left the country, except that the turban of the princess was not removed because of her royal status. Once in her new home, the princess carefully planted the mulberry seeds she had brought. As soon as the trees were large enough, she supervised the development of the silkworms as she formerly had in her original home. The breeding of silkworms and the production of silk is thought to have spread throughout India from that small beginning.

The second royal culprit was a renegade prince who went to Japan, where he became a citizen, in about A.D. 200. He is reported to have told representatives of the Japanese imperial family about silkworms and the manufacture of silk. He had taken with him some silkworm eggs, which were cultivated. However, the Japanese still lacked knowledge of how to weave the silk, and so the Japanese emperor sent some Koreans on a secret mission to China. They were to bring back information on weaving, but in lieu of it they brought back four Chinese maidens who were skilled in weaving techniques.

It is not known whether the young women went voluntarily or were taken forcibly, but upon their arrival in Japan they instructed members of the royal family and carefully selected noble families in the art of silk weaving. From that time on, silkworm breeding and silk weaving became an important Japanese industry. It is, in fact, credited with helping to establish Japan as a world power. The Japanese guarded the secret as carefully as the Chinese had and made every effort to prevent it from being obtained by other countries.

Another recorded theft of the Chinese silk secret involved the Emperor Justinian, A.D. 529 to 565, described as the last great ruler of the decaying Roman empire. His subjects had been importing raw silk from the East at enormous cost. The silk was manufactured into garments in his capital, Constantinople, which was for centuries a great trade center for silk textiles. In those days even the "common people" wore silk garments; the only limitation was that the color purple was reserved for the nobility. By royal decree it was not to be used by "ordinary citizens no matter how exalted their stations."

Justinian decided, probably because of the excessive cost of raw silk, to obtain the silk secret from China so that both the fiber and the fabric could be produced locally and so that the silk industry could be independent of the rest of the world. In about A.D. 550, he sent two monks to China to obtain employment in the silk-producing industry and remain there until they had obtained the secret of raw silk production. Both monks had previously lived in China and so were familiar with the country.

Justinian's industrial spies remained in China for two years and learned as much as they could about the silk secret. In about A.D.

552, when they made the long return journey to Constantinople, they could inform the emperor that the secret of silk was a worm. Also, they could show him some silkworm eggs they had smuggled out of China in their bamboo walking sticks.

Justinian recognized that the monks had risked their lives to obtain the secret and bring with them the silkworm eggs, and he is reported to have rewarded them amply for their theft. Well he might, for the secret was probably the most important one of the time. The silkworm eggs were cultivated, and the empire became a famous producer of silk for the markets of the Middle East and Europe. It enjoyed that distinction for about 600 years. Industrial espionage was an important factor in the success of an entire country.

THE INCREDIBLE ACTIVITIES OF ALFRED KRUPP

The central figure of a more up-to-date saga of industrial espionage was Alfred Krupp. From 1826 until his death in 1887, Krupp was head of the German industrial giant that bore his name. His organization developed such proficiency in the manufacture of heavy weapons that the name "Krupp" became synonymous with armaments. Krupp is credited with a maximum use of spies to keep abreast of developments throughout the world and to expand the markets for the broad range of Krupp products. He also used German embassy officials, and his agents are reputed to have offered bribes on a large scale to obtain information.

Krupp is reported to have personally engaged in industrial espionage in the 1830s. He journeyed to England to obtain information about new steel manufacturing methods. Although he probably imagined that he was performing well in the strange world of industrial espionage, he actually presented a ridiculous figure. He spoke very little English and attempted to conceal the fact that he was a German. He apparently did not deceive anyone, and he obtained little or no information of value.

FIDDLER FOLEY

Krupp was not the first head of an organization to be personally engaged in the collection of information through industrial espionage. In the seventeenth century a well-to-do English ironmaster named Foley, of Stourbridge in Worcestershire, wanted to know the methods of treating iron and steel then being practiced on the Continent.[2] Because he was an accomplished violinist, he decided to disguise himself as an itinerant minstrel.

Foley capitalized on his humor and musical ability while he wandered barefoot and ragged through Belgium, Germany, northern Italy, and Spain in his attempt to learn secret processes from the iron workers of these countries. He apparently made friends with master craftsmen and was able to get them to reveal their secrets. When he thought he had accomplished his mission, "Fiddler" Foley returned to England, only to discover that he did not have all of the necessary data. He returned a second time to the Continent in the same disguise and on that trip obtained all of the needed information.

When Foley's success as an industrial spy became known, the foreign ironmasters and guilds were greatly disturbed. The acquisition of their secrets would enable Foley to enter the Continental market as a dangerous competitor. In reprisal, agents of the ironmasters and the guilds were sent to England, where they made several attempts to assassinate Foley and destroy his ironworks. The attempts on his life and his property were not successful, and the industrial espionage efforts of Fiddler Foley were reported to be directly responsible for the fortune that Foley accumulated and passed on to his descendants.

THE GREAT ARIES CAPER

Probably the most celebrated industrial espionage case in the United States in recent times involved the exploits of a brilliant and daring scientist named Robert Aries, a naturalized U.S. citizen who was born in Sofia, Bulgaria. Aries was a distinguished, successful chemist with an impressive list of academic titles that included a doctorate in chemical engineering from Brooklyn Polytechnic Institute, a master of arts degree from the University of Minnesota, and a master of science degree from Yale. He was listed in *Who's Who*, had been granted more than 40 patents, had written widely for technical publications, and had been teaching at the Brooklyn Polytechnic Institute since he was 22 years of age. In addition, he lectured at the University of Geneva and was a respected consultant to the federal government as well as a number of corporations.

Merck & Co. originally became aware of Aries' activities as an industrial spy in the early 1960s. Merck had developed a drug, amprolium, that was remarkably successful in destroying a deadly poultry parasite called coccidia. The parasite had long been responsible for wholesale and worldwide losses of poultry flocks, and amprolium was developed by Merck after several years of research at a reported cost of $1.5 million. The potential yearly market was $20 million.

An application for a patent on amprolium was filed, and by early 1960 the drug was ready to be marketed. At about the same time, Merck officials learned, from an abstract of a paper that was to be presented at an Ottawa scientific meeting, that Aries was going to claim he had developed an anticoccidial drug that he called mepyrium and that seemed to be identical with amprolium. Also, Merck was told by the management of the French chemical firm Synorga, a company that Merck had acquired, that Aries had provided Synorga with all the technical and production data on mepyrium and had licensed the firm to manufacture the drug for $8,500. Merck further learned that Aries had licensed several other companies to manufacture the product; they included Sterling Drug Inc. in the United States, Burroughs Wellcome Company in England, and Hoffman-La Roche in Switzerland. Aries had reportedly been paid $100,000 for the licenses and also had royalty rights from which he might earn millions.

It seemed to Merck management most unlikely that Aries could have independently developed a formula that was virtually identical with theirs. An immediate investigation by the Merck security organization established that data in Synorga's files had indeed been taken from Merck files. Aries had evidently had access to Merck's research files since September 1959. The question was how he had gained the access. Although the answer was still not known, Merck filed suit against Aries for $7.8 million and obtained a restraining order against further disclosure of the amprolium secret. Before the order could be enforced, Aries fled to Europe.

The investigation continued, and every possible clue to the loss of the valuable trade secret was examined. Because it was obviously an inside job, the first assumption was that Aries' conspirator was a Merck employee who had access to the amprolium files. More than one hundred employees were in that category, but the investigators were unable to find anyone culpable among them. Then a Merck employee reported that a Merck engineer knew Aries well and had been a student of his at Brooklyn Polytechnic Institute in the late 1940s. At first, that did not seem to be too encouraging a lead, because the engineer not only had never worked on amprolium but had never even been shown any information related to it. However, a review of his employment records revealed that his handwriting was much like that of the handwritten copies of Merck documents then in the Synorga files.

The engineer had joined Merck in 1941 and was a key man in his division; in fact, the company considered him such a valuable employee that it had paid part of his master's degree tuition. When he was confronted with the evidence against him, he admitted that he had not only copied data from the amprolium project research files but had also taken samples of the compound. He had given both copies and samples to Aries. He admitted that Aries had offered to buy data from him over the years and that finally, in 1959, he furnished Aries with the amprolium information. Aries had recognized the formula to be extremely valuable. He told his former student that he would obtain worldwide patents and that he would give him 25 percent of the proceeds of the patents.

It was also learned that Aries had made a practice of attempting to obtain sensitive information from his students. He would assign class projects that involved written reports, and he would suggest that his students include interesting technical data about the everyday sensitive work in which they were engaged. It was obvious, his former students reflected, that he had more than an academic interest in their work, because he often remarked during his lectures on the potential value of the information they had available to them.

When the methods by which Aries had obtained information from Merck were publicized, other companies started investigations. It was found that Aries had used other former students to obtain valuable information. For example, he had received a formula for a new lubricating oil additive from Rohm & Haas Company and data on a device utilized in computers from a former student who was an employee of Sprague Electric Company.

Both Sprague and Rohm & Haas, in addition to Merck, filed damage suits against Aries, and in late 1964 judgments totaling more than $21 million were awarded to the three companies by a federal district court in Connecticut. Merck was awarded $6,637,499; Sprague, $8,600,000; and Rohm & Haas, $6,038,000. Because Aries was already in Europe, the judgments were levied against some

American companies he had organized. However, he left those companies practically insolvent, so the judgments turned out to have little or no value.

Aries has been the subject of criminal indictments in the United States, and both criminal and civil actions have been filed against him in France and Switzerland. At last report he was living in luxury somewhere in Europe and proclaiming himself the innocent victim of a vicious conspiracy and not an industrial spy. His activities do not seem to have changed much. In early 1972 he was busily filing trade names in various European countries to block their use by large U.S. companies doing foreign business. Typical was his filing a trade name similar to EXXON, which Standard Oil Co. (N.J.) is trying to establish as its worldwide trade name.[3] It remains to be seen how much that activity, which is legal if it is done without relying on espionage to obtain advance information on target names, will be worth to him.

MR. CREST AND MR. COLGATE

Another recent case demonstrates that scientific research data need not necessarily be involved to make company information extremely valuable.[4] This example of industrial espionage, which involved intrigue, an inept spy, and an element of comedy, revolved around the sale of the marketing plan for Crest toothpaste for the year 1964–65. The price was $20,000, which was a bargain because the owners estimated that the plan could be worth as much as $100 million to a competitor. The plan contained details of special promotion projects, and so a competitor in possession of the data could time his own promotions to saturate the market just in advance of the Crest promotions. That would make the plan worthless and cause both developmental and sales losses. The series of events, which started with a long-distance telephone call to an acquaintance in a rival company and ended up in a men's rest room at a major airport, were bizarre enough to be the basis of a movie or television thriller.

Eugene Mayfield, a 24-year-old graduate of Oberlin College, was hired by Procter & Gamble in Cincinnati, Ohio, as an executive trainee in 1962. He worked for two years as a writer in the advertising department and then resigned, in July 1964, to join another firm in Chicago. Before he left, however, he made copies of a 188-page document that contained the complete marketing strategy for the next year for Crest toothpaste. In November, from his home in Illinois, Mayfield called an acquaintance at Colgate-Palmolive Company in New York City and offered to sell the document. Colgate-Palmolive's Colgate toothpaste is, of course, a leading competitor of Procter & Gamble's Crest.

The Colgate employee reported the incident to representatives of his company, who in turn notified the Federal Bureau of Investigation. With the concurrence of the FBI, the Colgate management appointed a contact to negotiate with Mayfield for the purchase of the stolen data. Mayfield apparently felt that the modus operandi of a good spy should include a code name, so he instructed the Colgate man to refer to him as Mr. Crest. It was agreed that Mr. Colgate would meet Mr. Crest in a men's room in the TWA terminal at Kennedy International Airport in New York. At the agreed-upon time, Mr. Colgate went into a designated compartment. Mr. Crest went into an adjacent compartment and instructed Mr. Colgate to hand him $20,000 under the partition. The money was in a packet of marked bills. To foil pursuit, Mr. Crest also told Mr. Colgate to pass his trousers under the partition. Mr. Crest then dashed out of the rest room with the marked money—and was arrested by the FBI agents who were waiting for him.

Two federal statutes were involved in Mayfield's brief, amateurish excursion into the wonderland of industrial espionage. The first makes it a crime to utilize a telephone in interstate commerce with intent to defraud, and the second forbids the transportation of stolen property across state lines. Mayfield was charged with violation of the latter. Although the statute provides for imprisonment up to ten years and a fine up to $10,000, Mayfield, who pleaded guilty, was sentenced to imprisonment for two years on August 5, 1965. The sentence was suspended.

SUMMARY OF CASES

The thefts of the Chinese silk secret demonstrate that people have for thousands of years been tempted to steal the trade secrets of others. The sinister events and human suffering that must have occurred over the centuries as individuals attempted to obtain the secret can easily be imagined. And the fact that on two of the occasions the secret was lost as the result of an inside job shows that industrial spying has not really changed since the recording of history.

The four more recent cases are not classic examples of industrial espionage. Krupp, Foley, Aries, and Mayfield had one thing in common: they were unsuccessful industrial spies. Krupp was so ridiculous that it is obvious why he was not successful in his personal efforts. Fiddler Foley was able to obtain the information he wanted and to put it to good use, but his espionage activities later became known. Then his life was threatened and attempts were made to destroy his business. Although he survived the threats, the fact that his operation became known made it unsuccessful.

Aries may appear to have been very successful in his efforts; he is reported to have become wealthy, and he seems to have escaped punishment. However, he was unsuccessful because he was caught, because his activities became known, and because he has been forced to live in exile and in constant danger of being incarcerated. Mayfield was unsuccessful because he was simply a naive, inept young man who conducted himself in such a foolish manner that he was caught and punished immediately.

The reader might gain the impression that such characters are typical of persons engaged in industrial espionage and so there is really little to fear. Not so. As we mentioned earlier, the successful industrial spy is the one who obtains the information he wants, delivers it to a buyer for a high price, and then fades away and is never caught. More than that, the victim of a truly successful operation will never know that the information has been stolen. An operator of that type is the most insidious one and the one who is most difficult to cope with. In the remainder of this book the reader is instructed in how to best protect himself and his organization from that type of industrial spy.

REFERENCES

1. William F. Leggett, *The Story of Silk* (New York: J. J. Little and Ives Company, 1949).
2. Richard Wilmer Rowan, *The History of Secret Service* (New York: The Literary Guild of North America Inc., 1937).
3. "Trademarks: Identity Crisis," *Newsweek*, July 3, 1972, p. 63.
4. *United States* v. *Mayfield*, 65 CR 143 (E.D.N.Y. 1965).

2

WHAT IS
WORTH STEALING?

Information, it has been estimated, accounts for 75 percent of the success of any executive. From the moment the average executive enters his workplace each day, he is mainly concerned with collecting, evaluating, and acting upon it. Because of its importance, it should be regarded as a valuable asset and protected in the same way that money would be. At the start of planning an espionage prevention program, therefore, the information about the organization that is most likely to be an industrial espionage target must be determined. It is that type of information that will be discussed in the remainder of this chapter.

The organizations most vulnerable to industrial espionage are generally thought to be those involved with electronics, automobiles, chemicals, oil, toys, and drugs, but loss of information by theft is not limited to them. Thanks to the economic struggle for survival, every organization, regardless of size or type of activity, has information the loss of which could result in sufficient harm to force it out of business. For that reason some of the information that should be protected will be discussed without mention of the type of organization most likely to have it. Management representatives often naively assume that their organizations do not have any data worth protecting. Many readers who have adopted that attitude will be surprised to find that they actually have a great deal of information that could be of value to a competitor.

PREVENTION—THE KEY

The basis for any industrial espionage prevention program is protection of information. A Chinese proverb attributed to Confucius is that a secret is your servant if you keep it but your master if you lose it. Someone else has compared information to perfume—once it has escaped from the bottle, it can never be recaptured. Legal recourse may be available if data are improperly obtained, but experience indicates that it is not only costly but ineffective. Once the information has been lost, the enterprise can never be completely compensated for the damage done. So although legal actions will be discussed in later chapters, loss prevention will be constantly stressed as the key to the protection of information.

In general, the information subject to industrial espionage is any kind that would improve the competitive posture of a rival. That is made obvious by the nature of competition, which Webster defines as the effort of two or more parties to secure the custom of a third party by the offer of the most favorable terms. The two main elements of commercial competition are (1) the production of goods or services and (2) the locating and retention of customers who will buy the goods or services produced. Any information that involves either of those two elements is a valuable asset that requires protection. It is not limited to data that bear directly on production and sale of items. It also includes related information about the organization that would be of value to a competitor.

RESEARCH AND DEVELOPMENT INFORMATION

The production of sensitive data usually begins with an idea. The idea may involve a new product or service or simply a concept that will place the individual or organization in an improved competitive position in the marketplace. Protection of the basic idea is important because the idea itself might be of more value to a competitor than its end product.

It is not necessary that the idea be developed by much work or experimentation; it might be only a flash of genius. The key to its importance is the value it would have to a competitor. Usually, however, an idea cannot be effectively applied until it has been researched, so research and development are commonly the next steps in the creation of a product or a service. Research may be a relatively simple process such as the collection of statistical data by which the validity of the basic idea can be determined and on which a course of action can be developed to exploit the idea. Research and development, then, are not necessarily confined to a laboratory or the research and development facilities that may be found in large organizations. Every enterprise, regardless of size, will conduct research and development, although they may not be thought of as such.

Research will usually result in a drawing, formula, design, plan, pattern, specimen, process, or course of action that will be the basis of the development and sale of a product or service. At this stage information will often be of particular value to a competitor and its loss of extraordinary cost to the developer. The Aries case discussed in Chapter 1 illustrates both the great loss that can be sustained and the value that can be gained as a result of the theft of an item in the research stage.

Another case that illustrates the value of research information to a competitor involved 3M Company and Technical Tape, Inc.[1] At a cost of more than $3 million, 3M had solved a problem that was still causing its competitors difficulty; it had developed a formula that was used to manufacture a cellophane tape that would peel cleanly off the roll.

The chief executive officer of Technical Tape, Inc., who resided in New Rochelle, New York, decided in early 1951 that he would attempt to obtain the 3M secret process. He directed the company salesman in the St. Paul territory to contact Frederick Beyer, a 3M employee who had access to 3M company trade secrets. Beyer was invited by the Technical Tape representative to go to New Rochelle to meet with the company's chief executive officer to discuss employment. In March 1951 the Technical Tape chief executive offered Beyer about a third more salary than he was earning at 3M with the understanding that he would obtain information regarding 3M formulas and processes.

Beyer accepted both the offer of employment and the condition; he would remain on the 3M payroll only until he had obtained the information wanted by Technical Tape. At a meeting in the Technical Tape plant in the Bronx, New York, with representatives of the company, he discussed 3M processes and techniques. When he left to return to his job at 3M, he took with him a handwritten list of questions asked by the chief executive officer and a vice-president of Technical Tape; and later he sent the vice-president answers to all the questions. He also copied from 3M files ten formulas of chemicals used in the manufacture of tape and gave them to the Technical Tape representative.

Beyer terminated his employment with 3M in May 1951. He informed 3M representatives, at the suggestion of the chief executive of Technical Tape, that he was going to work for a company not in the same business as 3M. For that reason, it was a year before 3M management representatives could determine how the valuable information had been obtained.

In defending the ensuing action brought by 3M, Technical Tape alleged that the secret obtained from 3M could have been discovered by a diligent search of expired patents and other information that was available. The court held that to be immaterial: the secret formula had not been obtained through either the literature or laboratory experiments; it had been obtained by illegitimately approaching Beyer and inducing him to breach an agreement he had with 3M not to disclose confidential information. Damages and costs were awarded to 3M, and Technical Tape was permanently enjoined from further use of the secret.

An idea need not have been effectuated to be of value. After research, it may be found not to have any merit, or the organization may decide against utilizing it. It may still be in need of protection because knowledge that it was considered and discarded might be of great value to a competitor. For example, one company spent a large amount of time developing the technology for a product for which a large demand was expected,[2] but at the last minute top management decided against introduction because the organization was too busy with other product lines. The day after the decision was made, the president received a call from a marketing man in a rival company. The competitor knew of the decision to postpone the introduction of the product, and the marketing man told the president he was disappointed because his company had intended to offer a tie-in sale. Information about the new product had evidently been of value: the competitor had made plans to utilize it. Knowledge of the cancellation was also of value, because representatives of the rival company could then revise their plans accordingly.

PRODUCTION INFORMATION

The next step after the research and development stage is the production of the product or the preparation of the service for sale. At that stage of the development any number of items of information about an idea may be of value to a competitor. If a machine is involved in the production, information regarding its design and operation may be of value. Extraordinary advances in the development of machines have been made, and in recent years electronic and chemical technologies have been added to mechanical ones.

On the other hand, a machine may be a simple contrivance; but if it produces a special product, it should be protected so that a competitor cannot gain an advantage by learning about it. Manufacturers of fishhooks carefully guard the secrets of how their machines apply the barb, form the point, and shape the eye of the hook. The companies design and manufacture the relatively simple machines they use and allow only those who operate the machines to have access to the production area.

The problem of the loss of secret information about machines is demonstrated by a court action brought by the Monsanto Company.[3] At considerable expense the company had developed an electric furnace to produce elemental phosphorus, and it pursued the policy of treating the design as secret. Charles M. Miller, who had worked for Monsanto from 1942 to 1954, was familiar with the technical details of the furnace and also knew that Monsanto regarded the whole operation as a trade secret. He had become a consultant after leaving Monsanto, and in 1956 he was engaged by Central Farmers Fertilizer Corporation. Later he became an employee of that company.

Central Farmers was interested in the design and construction of an electric furnace for the production of elemental phosphorus, and Miller worked on the project. Before he left Monsanto, he had obtained company engineering drawings, cost data, an operations manual, and other technical information related to the Monsanto furnace. Fortunately, Monsanto management representatives became aware that they had lost the furnace trade secret when Central Farmers began to construct a plant with an electric furnace similar to the one Monsanto had developed. The case was decided in Monsanto's favor because it was clearly shown that Miller had improperly obtained trade secrets for a competitor.

Still another aspect of the production process that may be of value to a competitor is the end product. The information of value may concern the ingredients, their source of supply, or their amounts and the way in which they are blended in the production process. A product secret that has been protected by one family for centuries is that of the Zildjian cymbal.[4] The method of producing and treating a certain alloy to produce the cymbal, regarded by musicians as the best in the world, was discovered by the Zildjians in Constantinople in 1623. Today a factory operated by family members in Quincy, Massachusetts, produces cymbals under very strict security controls to insure that the integrity of the secrets of the product is maintained.

Two alcoholic beverages—Drambuie liqueur and Tribuno vermouth—are other well-known examples of secret products.[5] The formula of each is known only to one person in the producing company. Usually, such a formula is kept in a bank vault and a carefully controlled access list is maintained.

Know-how is also an important aspect of production and could therefore be of value to a competitor. It has been defined as [6]

Factual knowledge not capable of precise, separate description but which, when used in an accumulated form, after being acquired as the result of trial and error, gives to the one acquiring it an ability to produce something which he otherwise would not have known how to produce with the same accuracy and precision found necessary for commercial success.

Trial and error, then, is the key to know-how, which could involve information concerning the improvement of methods already known. And although it is usually thought of as closely related to the production phase, it might also consist of techniques developed in the general business or marketing areas. For example, it might represent knowledge of mistakes to be avoided, which would be of enormous value to a competitor interested in entering a particular field of endeavor. Know-how is often developed at great cost to the enterprise.

A variety of other aspects of production might involve information of interest to a competitor: manufacturing status, problems, costs, and schedules. Production, as discussed in this section, is not limited to hardware items. Any service organization will have information related to the preparation or production of the service that would be valuable to a competitor. A good example of such an organization is an advertising agency. The following excerpt from a feature story in the *Los Angeles Times* reflects the attention that one advertising agency is giving to information that might benefit another.[7]

The cloak-and-dagger world of internal security is beginning to make inroads on Madison Avenue's freewheeling advertising business. It's not so surprising when you consider that advertising agencies deal in the most fragile of commodities: ideas. Actually, business in general is placing increasing emphasis on security, both against industrial espionage and acts of terror, theft and vandalism. Yet few agencies, even some of the biggest, have formal security programs. One that does is Ogilvy & Mather, Inc., a major New York agency. Says Ed San Luis, manager of administrative systems:

"It's rare when you hear of secrets being lost from one agency to another—but it does happen. Statistics are not compiled on this sort of thing, mainly because incidents are hushed up by companies. That's why security has to be a before-the-fact thing; afterwards, nobody wants to talk about it. There are many cases in which an agency sees a competitor come out with a campaign similar to its own. Its executives wonder if somehow the word got out. It's hard to keep advertising secrets because there are so many external people involved."

Another example of a nonhardware product is a computer program. Information about computer software is becoming increasingly important. A software program may be developed at great cost and be a correspondingly valuable product. Because computer security is so important, Chapter 8 is devoted to it.

MARKETING INFORMATION

The difference between success and failure of a product or a service may depend on one important factor—getting to the marketplace before the competition. In recent years, in fact, the emphasis in product exploitation has shifted from technology to marketing. There was a time when a company that had an item that would perform better had no problem in selling it, but now the consumer is made to believe that an item is better by promotion and marketing. It has been estimated that about 12,000 new drug and grocery products are released each year. That is more than twice the figure of ten years ago. As a result, marketing data can often be more important to a competitor than the other areas of information already discussed in this chapter.

Lead time—the amount of time in the market the introducer of a product has before competitors can produce copies—is often an important factor in the success of an item. A few years ago competitors were less adept at copying than they are now and so lead time could be routinely expected. Today, because of improved technology, competitors are able to react and enter the market more quickly with lower-cost copies. Also, a company must now spend a great deal more to introduce a new product or service than it did a few years ago. For that reason it will make every effort during the marketing stage to avoid an ambivalent consumer reaction. According to an article in a national publication: [8]

Of the roughly 100,000 soaps, food snacks, and other products that will bomb in supermarkets during the 1970s—out of a total of 120,000 introductions—the advertising expense alone will add up to an estimated $5 billion for those being test-marketed and $7 billion or $8 billion for those introduced nationally.

In only six months, Gillette's Trac II razor went from an unnamed, handmade prototype to a finished product with enough production capacity—more than one million blade cartridges a year—for a national kickoff by World Series time. For the initial big push, Gillette spent a hefty $4 million on advertising and sales promotion, and between last October and next October will lay out another $6 million on advertising and promotion.

Another example of the high cost of marketing is the effort made by General Foods Corporation to introduce a cereal that contained freeze-dried fruit. The product failed, but it has been estimated that it cost the company $15 million—$10 million for promotion and $5 million for a plant.

The value companies place on marketing information is illustrated by the Mayfield case discussed in Chapter 1—Mr. Crest and Mr. Colgate—which involved the theft of a complete company marketing plan. Marketing information of value to a competitor may include a wide variety of items: advertising plans; special promotion and strategy plans; test-market plans and results; prices, sales data, and future plans; and premiums, coupons, or other promotional lures. Protecting information about a product or service after it has been introduced is also important. Some items of marketing information of particular value to a rival are sales statistics, customer information, profit data, plans for sales campaigns, and exploitation of territories.

Frequently, customer relations constitute an organization's most valuable asset. Of particular value to competitors are the customer lists that every competing organization, regardless of size or activity, must have. The lists of great value are those that are unavailable from public sources and have been developed by the owning enterprises over the years at great expense. Many cases that involve customer lists that have been unlawfully obtained by competitors have been in the courts over the years. Often a list is stolen by a terminating employee, who then turns it over to a competitor or uses it in his new employment or in a business that he starts. Other information of great potential value to a competitor includes background data on customers, selling policies as they relate to each customer, credit data, and sales amounts. A 1970 case involved three employees of the *Encyclopaedia Britannica* who obtained and sold the company's confidential list of customers. The 800,000 names, which were contained in seven reels of magnetic computer tape, were sold for direct mail solicitation.

It should be noted that an individual who brings a customer list into a company is usually held by the courts to be the owner of that list unless he unequivocally indicates an intent to transfer ownership to the company. If he later leaves, he cannot usually be prevented from taking the list with him unless a contrary arrangement has been made.

MISCELLANEOUS BUSINESS AND OTHER INFORMATION

In addition to information in the three areas already discussed—research and development, production, and marketing—every organization will have other information that would be valuable to competitors. Long-range business plans represent data that might be of value to a competitor, who might use the information to take actions that benefit his organization and at the same time have an adverse effect on the company that originated the plans.

Other business information is of potential interest to competitors; examples are accounting records having to do with costs, general financial affairs, and the commitment of resources. The importance of cost accounting information submitted by defense contractors has been recognized by the Department of Defense. In February 1972, the Cost Accounting Standards Board transmitted to Congress for approval a list of proposed standards, rules, and regulations relating to cost accounting by defense contractors. It was recognized in a section in the standards that information in any disclosure statements that were submitted should not be made public, because competitors not required to submit such cost information might obtain a competitive advantage if they had access to it. The section specifies that a disclosure statement will not be made public in any case when a company or division files a statement conditional on the government's agreement to treat it as confidential information.

Information relating to reorganization, methods of doing business, business opportunities being considered, executive promotion, reassignment, or compensation, closing or relocation of plants or other facilities, expansion plans to include mergers and acquisitions, and prospective litigation should also be considered a potential target for a rival.

A case involving the misappropriation of a detailed business plan was decided in January 1972.[9] The plan included all of the forms, information, and techniques for formulating, promoting, financing, and selling contracts for "prepaid" or "pre-need" funeral services in the continuous operation of a mortician's business that would give the user a marked advantage.

The defense alleged that the selling of pre-need funerals was common practice and that many elements of the particular plan had in fact been taken from the plans of others. The court held, however, that although the defendant, Berkeley L. Bunker, might have collected the same information from sources available to the public, substantial time, thought, and money had in fact been expended by the plaintiff in collecting and testing the data as well as the techniques that provided the substance and detail of the business plan. The court noted further that substantial effort would be required to assemble the detailed elements of the plan from public sources. Accordingly, damages were awarded to Larry E. Clark, who had developed the business plans.

REFERENCES

1. *3M Company* v. *Technical Tape, Inc. et al.,* 192 N.Y.S. 2d 102, 123 U.S.P.Q. 96 (1959).
2. John Perham, "The Great Game of Corporate Espionage," *Dun's,* October 1970, p. 33.
3. *Monsanto Company* v. *Miller,* 118 U.S.P.Q. 74 (1958).
4. William Bowen, "Who Owns What's in Your Head," *Fortune,* July 1964, p. 177.

5. Victor Block, "When a Secret Is Better Than a Patent," *Popular Mechanics*, November 1965, p. 110.
6. *Myclex Corporation* v. *Pemco Corporation*, 68 U.S.P.Q. 317 (1946).
7. Martin Rossman, "Agencies Get Hush-Hush on Ideas They Intend to Blare," *Los Angeles Times*, February 7, 1972. Used by permission.
8. "New Products: The Push Is on Marketing," *Business Week*, March 4, 1972, p. 72. Used by permission.
9. *Clark* v. *Bunker*, 172 U.S.P.Q. 420 (1972).

3

CONFIDENTIAL INFORMATION
AND TRADE SECRETS—
LAW AND FACT

Many kinds of information are regarded as confidential, but not all of them can strictly be classified as trade secrets. Often, of course, it will be just as important in an individual situation to protect confidential information that is not a trade secret as that which is. And because the law provides different rights and remedies in the different circumstances, failure to perceive the distinction could result in a compromise or disclosure for which no adequate remedy would exist.

The purpose of this chapter is to trace the general lines followed by the law, common and statutory, in dealing with confidential information. The material presented is not intended to be a definitive legal treatise or to provide a basis for specific case research. There are excellent resources already available to serve those purposes.[1] To put the practical do's and don'ts of sensitive information protection in a legal context is more our aim here. In the final development of either a prosecution of or a defense to an alleged misuse or theft of confidential information, competent counsel must always be consulted. However, in the early days of a protection program it will often be the enterprise executive himself who must be alert to and who must deal properly with problems involving the sensitive information of his organization. It is essential to have clear ideas about some basic legal considerations in order to do so.

Situations involving confidential information may fall into the ambit of the statutory criminal law, federal and state, the statutory civil law, again in both jurisdictions, or the common law. With notable exceptions, the common law is generally the law of the state even though it is sometimes applied in federal tribunals, as when diversity of citizenship brings litigants into the federal courts. The most active of the areas mentioned has been that involving the common law, and so it is appropriate to begin the discussion there.

THE COMMON LAW

The common law provides two classical approaches to dealing with sensitive information. One, the so-called property concept, regards the information, or quite often the physical item embodying or manifesting it, as having an independent value as property in much the same way that hard assets like supplies, tools, and finished products have, provided it amounts to what will later be defined as a trade secret. That approach holds that the owner of trade secret information is the owner of property and is entitled to all the safeguards and remedies the law would provide him for the protection of any other property. Such rights include the right to sue in damages for the loss or destruction of the property, the right to recover profits unjustly made by another while using the property without authorization, the right to restrain another from completing some threatened use that would result in serious harm to the true owner, and the right simply to retain exclusive use and dominion over the property, at least to the extent that others have not independently and without any overreaching or improper conduct developed identical or similar information themselves.

The second main line of common law on sensitive information imposes duties upon certain classes of persons other than the owner not to use or divulge the information without the owner's consent. Such persons are those who are in special positions of trust and confidence—fiduciaries in legal parlance—or those who have entered into express or implied agreements by which they bind themselves to protect it. An example of the former class would be the ordinary employee who, whether he wishes to be or not, is held to a duty not to divulge or use his employer's confidential information without authorization. In that class too, however, the law generally requires that the information be able to meet customary tests as a trade secret. Other information, however confidential the employer may regard it as being, will not be protected against use or disclosure by mere fact of the employment relationship. The distinction is critical, and more will be said about it later.

In both the property concept and the duty concept situations, then, the common law has generally required that, in the absence of specific agreement, the information protected qualify as a trade secret. The distinction between trade secret and other confidential data is therefore central to securing common law protection. There are literally dozens of definitions to be found in the case law of various jurisdictions, and no definition is binding outside the state whose courts developed it. However, a consensus of the elements most often agreed to be required may be found in the *Restatement of the Law of Torts*. The *Restatement* is a work of legal scholarship sponsored by the American Law Institute that seeks to identify, in many areas of the common law, the generally agreed elements of a right, a wrong, or a relationship. The *Restatement* definition of trade secret is as follows.[2]

Definition of a trade secret. A trade secret may consist of any formula, pattern, device or compilation of information which is used in one's business and which gives him an opportunity to gain an advantage over competitors who do not know or use it. It may be a formula for a chemical compound, a process of manufacturing, treating or preserving materials, a pattern for a machine or other device, or a list of customers. It differs from other secret information in a business [defined in another section of the *Restatement of Torts*] in that it is not simply information as to single or ephemeral events in the conduct of the business, as, for example, the amount or other terms of a secret bid for a contract or the salary of certain employees, or the security investments made or contemplated, or the date fixed for the announcement of a new policy or for bringing out a new model or the like. A trade secret is a process or device for continuous use in the operation of the business. Generally it relates to the production of goods, as, for example, a machine or formula for the production of an article. It may, however, relate to the sale of goods or to other operations in the business, such as a code for determining discounts, rebates or other concessions in a price list or catalogue, or a list of specialized customers, or a method of bookkeeping or other office managements.

Secrecy. The subject matter of a trade secret must be secret. Matters of public knowledge or of general knowledge in an industry cannot be appropriated by one as his secret. Matters which are completely disclosed by the goods one markets cannot be his secret. Substantially, a trade secret is known only in the particular business in which it is used. It is not requisite that only the proprietor of the business know it. He may, without losing his protection, communicate it to employees involved in its use. He may likewise communicate it to others pledged to secrecy. Others may know of it independently, as, for example, when they have discovered the process or formula by independent invention, and are keeping it secret. Nevertheless, a substantial element of secrecy must exist, so that, except by use of improper means, there would be difficulty in acquiring the information. An exact definition of a trade secret is not possible. Some factors to be considered in determining whether given information is one's trade secret are: (1) the extent to which the information is known outside of his business; (2) the extent to which it is known by employees and others involved in his business; (3) the extent of measures taken by him to guard the secrecy of the information; (4) the value of the information to him and to his competitors; (5) the amount of effort or money expended by him in developing the information; (6) the ease or difficulty with which the information could be properly acquired or duplicated by others.

Although the differing of the case law with the jurisdiction makes the *Restatement* definition only one of the many that are in fact used, we can confine our attention to it for the purposes of this general discussion. Note that it requires three basic things of the information: (1) that it be of competitive advantage, (2) that it be secret, and (3) that it be used in the business of the owner. None of the requirements is absolute; that is, it is not essential that the competitive advantage be a present one or that the secrecy be so complete that only the owner knows the information or that the secret be used constantly. However, to the degree to which the information in question does not conform to a strict definition of each of the requirements, there is an increased likelihood that a court may not find it to be a trade secret. Note also that the information must be more than a confidence affecting the business; it must be used (or at least developed and intended for future use) on a regular basis.

The illustrations in the *Restatement* of the kinds of data that do not qualify as trade secrets, such as salary, contract bid, investment portfolio makeup, and new model date, do not include all the kinds of information that may not be granted trade secret protection. Among others that have been denied such status are merchandising methods, sales training methods, cost, pricing and profit information, sources of supply, and general business methods. To make the matter more difficult, some jurisdictions have found exactly to the contrary and have held such items to be protectable as trade secrets. It should be clear, then, that the question in a given case will hinge on the precise facts of that case. In general, however, the *Restatement* guides are good starting points.

If the information protectable as a trade secret may be an elusive concept, the degree of secrecy required is a little easier to determine. Nowhere is absolute secrecy required. At the same time, the cardinal principle to be observed is that unprotected or unrestricted disclosure puts what was or might have been a trade secret into the public domain and ends all enforceable proprietary interest thereafter. So the trade secret owner may make some disclosures, but he must be certain that none of them are unprotected or unrestricted. Practically, that means that all persons to whom disclosure is made either must be under a preexisting legal duty to protect the information (the fiduciary discussed a bit earlier) or must be made to accept such a duty by contract, express or implied. A later chapter deals in detail with the types of agreements that can be used to create enforceable rights against persons who receive disclosures. At this point it is necessary to establish only that the trade secret owner has the responsibility, at the outset, to assure that the persons to whom he makes protected or limited disclosures are under a legal duty, through status or agreement, to protect the information disclosed.

If the kind of information involved does not qualify as a trade secret, then the enterprise controlling the information has only one option: the agreement to protect. Reliance upon fiduciary status alone may not suffice for information that does not constitute a trade secret. Moreover, even when the information is protected by agreement, improper disclosure may really be irremediable. For example, if an employee learns of a proposed property acquisition by his employer and, before the employer can close the matter, makes his knowledge public, the price of the property may escalate and perhaps cause the acquisition to fall through. Even if the employee were also bound under a formal agreement not to divulge the fact and even if, in the given case, the acquisition could be demonstrated to be so important or central to operations as to be a trade secret (such as prospect sites on geophysical maps of an oil

company), the disclosure might cost the enterprise a great deal
more than it could ever recover from the employee. If in such a
case the firm's interest in the acquisition became public knowledge
and potential vendors were determined to raise the asking price,
the enterprise would be unable to compel them not to. If the
enterprise proved that a particular vendor knew that the employee was disclosing what probably was a trade secret, it could
prevent the employee concerned and the vendor from profiting
unjustly, but it could not compel the vendor to sell at a fair price
or, indeed, at any price. Its opportunity might thus be lost without
effective remedy.

The crucial points so far established are these: (1) Trade secret
information is entitled by law to more protection than other kinds
of confidential information. (2) For trade secret protection it will
be necessary to prove all the essential elements, namely, secrecy,
value to the enterprise, and use in the business of the enterprise.
(3) The quantum of proof of those elements that is needed varies
with the jurisdiction. (4) For an established trade secret, the proprietor may achieve his protection through the fiduciary status of
the disclosee or through agreement with the disclosee. (5) For
other confidential information, the enterprise is mostly left to de-
manding agreement to protect it before making disclosure.

But there is a tendency operating to enhance protection of sensitive data. It is the disposition of courts to find against persons in
bad faith, even when the precise status of the confidential information involved may be in some doubt. If the provable facts show
that the defendant in a misuse-of-information case had no permission or authority to use or disclose the data and that he knew or
clearly ought to have known that the information involved was the
property of the plaintiff, was regarded by the plaintiff as valuable,
was preserved secret to the maximum extent consistent with the
nature of the plaintiff's business, and would result in some measur-
able harm to the plaintiff if allowed to be used by the defendant
or others involved with him, there is a good probability that the
court will seek to protect the information and restrain the damaging conduct.

Thus, whether or not express or implied protective covenants
exist, it is of great practical value to the owner of confidential information that he be able to show the positive steps he took to
prevent its unauthorized divulgence. Without proof of diligent care

the best-defined trade secrets can be lost. With proof of diligent care, protection may be afforded to material that qualifies only marginally as trade secret.

Later chapters offer descriptions and checklists of the things that should be done to establish a program of diligent care. Of course, the primary reason for taking most protective measures is to prevent the actual unauthorized disclosure, not to make a convincing record for litigation. However, as the myriad cases show, litigation is often the last stop in many confidential information situations. It may also be important, therefore, to make that record. There is no way to know in the beginning.

REMEDIES UNDER COMMON LAW

The remedies available to a proprietor of confidential information under common law depend to some degree on the theory relied upon, that is, whether a contract (agreement) theory or a tort (violation of a fiduciary duty or other intentional injury) is alleged. In general, however, two forms of relief are available. The first is a restraint on use of information already disclosed or on its further disclosure. That is usually imposed by a temporary restraining order or by a temporary or permanent injunction. Each of those devices is an order of the court that directs specific persons to refrain from certain conduct. A former employee might be enjoined, for example, from revealing to a new employer the trade secrets of his former employer. Also, the new employer might be enjoined from using trade secrets of the former employer that had already been revealed to him and that he knew or ought to have known were trade secrets.

In some situations the employee may be enjoined from even taking employment with the new employer. Because that recourse is such a drastic remedy under the U.S. system, it is not readily granted by the courts. In the absence of a formal agreement by the employee not to work for competitors (discussed in Chapter 10), a court will generally not enjoin such employment unless the evidence is quite persuasive that it would inevitably and necessarily result in unauthorized disclosures of trade secrets. There is an interesting case in which the sensitive new employment was permitted but the employee was enjoined from disclosing trade secrets of his former employer in the new job.[3] The problem of policing such a delicate situation is most difficult, and the slow erosion of the secret, even with both new employer and employee in the best possible conscious good faith, is a real probability.

It is important to remember that the law does not afford protection against use of alleged trade secret information by a person who acquires it innocently and without notice, nor does it protect such information in the hands of an innocent acquirer, even after notice, if before the notice the acquirer expended significant value or significantly changed his position in reliance upon his right to use the data. And in any civil case in which wrongdoing is alleged, the person who alleges it must prove the charge by a preponderance of the evidence. The chances that a plaintiff will prevail in alleging improper disclosure or use of trade secrets depend upon timely knowledge or awareness that such use or disclosure may be occurring and convincing evidence to prove the improper conduct charged.

One technique that the proprietor of a trade secret often uses to help establish his ownership in the event of unlawful use of the information by a competitor is the inclusion of unnecessary or deliberately irrelevant data. For example, a process secret might include one or more steps that do not contribute anything to the end result. A formulation might include an inert ingredient that imparts no property to the final product. By the inclusion of such data, plus adequate documentation of the action, it may be possible at a later time to establish theft, or at least unauthorized use, if the same unnecessary step or ingredient is found in a competitor's effort.

This same technique is very popular with owners of mailing lists, certain kinds of which have been held to be trade secrets. The list owner includes fictitious names that are perhaps even coded and changed at timed intervals. Some or all of them will have addresses that in reality are drops or are otherwise under the control of the owner of the list. By inspecting the various mailings he receives, the list owner can determine whether they appear to be based on his trade secret lists. If they are, he at least has the identity of the mailer, his proof of ownership, and his precaution of salting the list with which to commence an action.

It is not difficult to see how such proof would impress a court. Even if the defendant alleged that the list could easily have been compiled independently and even if he showed that he was in a logical position to have so compiled it because of other contacts with or interests in the general population making up the list, the convincing evidence that he instead took the plaintiff's list—evidence supplied by the salted names, which could hardly be explained in any other way—would result in most courts deciding the matter on the palpable indications of defendant's bad faith without hair-splitting over whether the list or process or formulation was a true trade secret. Such practical self-help should be part of the security program of every enterprise with confidential data to protect.

In addition to preventing new or additional harm by requesting an injunction, the plaintiff in a civil theft-of-information case may seek money damages. Generally they are measured either by the provable cost or loss of profit suffered by the plaintiff or by the unjust gain enjoyed by the defendant through his misconduct. A very practical doctrine has been borrowed from patent law and developed over the years to help ascertain such amounts in certain cases. That is the doctrine that the amount of damages for use of a trade secret, if not clearly provable by other formulas, can be arrived at by establishing the royalty that would have been paid willingly by the defendant for a license granted willingly by the plaintiff. There will be situations in which that approach is not satisfactory either at all or as a total remedy. However, in cases in which the amount of compensation to the plaintiff is the only real sticking point, it has merit. Some litigation can be terminated by voluntary settlement by utilizing that approach.

FEDERAL CIVIL LAW

There is a growing body of statute law regulating the use of confidential information. Some of it is criminal and will be found at both state and federal levels. Some is civil and touches on the antitrust, unfair competition, and restraint-of-trade aspects of trade secrets. As to the latter, there is a broad rule that any agreement or practice that tends to lessen competition or restrain trade will be viewed with suspicion and that any such agreement or practice that has substantially serious impact will be held unlawful.

Both the people in general (through government intervention in court or in the administrative agencies) and private aggrieved parties have varying remedies against restraint of trade. A company, for example, might restrict its distributors, franchisees, or licensees in such a way regarding the use, price, or further disclosure of an item as to prevent a competitor from sharing in the market. If such a practice is confirmed through litigation, the prevailing aggrieved competitor may recover treble damages from the company practicing the restraint. However, if a bona fide trade secret is what is sought to be protected by restraints, much more restrictive conduct is permitted the proprietor of the secret.

One of the risks a large firm in a dominant market position takes when it seeks to impose restrictions upon licensees and others in the use of trade secret information is that in a subsequent litigation the information will be held not to be a trade secret and the company will thus be exposed to antitrust penalties for the restrictions it has imposed. The unfavorable result could come directly through attack by a competitor who files an antitrust action against the proprietor company or indirectly by way of an affirmative defense raised by a defendant who has gained unauthorized access to the plaintiff's alleged trade secret, is using it, and is sued by the owner.

Note that restrictions imposed by way of restraint of trade, such as price maintenance, limited use or territory, and nondisclosure, are not the same as the restrictions imposed by the proprietor of a trade secret (or other confidential information) who seeks only to protect his right to continue *his own* use of the secret. Thus, when a manufacturer requires a supplier of materials or components who must have access to the manufacturer's trade secret information to agree not to disclose that information to third parties, such a restriction is not restraint of trade, because the owner of the secret has the right not to permit others to use it.

Also, when an employer seeks to prevent a former employee from disclosing his trade secrets, he is not acting in restraint of trade. Even if the employee had executed a noncompetitive covenant (discussed in Chapter 10) reasonable as to term, territory, and subject matter, which would be in partial restraint of trade, it would still be generally allowable as a necessary corollary to the trade secret owner's right to protect his secret. Several states have expressly prohibited such noncompetitive covenants by statute, but there is respectable argument that a trade secret owner's rights in certain situations may be protectable only by contract, that those rights are at least equal to the public policy goal of preventing restraint of trade, and that injunction against a former employee taking some or any employment with a competitive new employer might still be available, despite the statutes, because to hold otherwise would be to run afoul of the constitutional right of contract.

FEDERAL CRIMINAL LAW

Because the act of industrial espionage is a conscious one involving the intent requisite for it to constitute a crime of larceny, it is especially appropriate to understand and invoke criminal sanctions in dealing with it. The chief federal sanction is found in the Federal Stolen Property Act, 18 United States Code 2314. The section provides:

Whoever transports in interstate or foreign commerce any goods, wares, merchandise, securities or money of the value of $5,000 or more, knowing the same to have been stolen, converted or taken by fraud . . . shall be fined not more than $10,000 or imprisoned for not more than ten years, or both.

There are two major difficulties with the federal Act. First, the definition of goods, wares, and merchandise has not been resolved to include the purely intellectual appropriation of another's trade secret for later use. The question is thus still open whether, if an employee or other person with lawful access to a trade secret used in interstate commerce carries it away in his head and later sells it to a competitor, such conduct would be violative of the section. Other cases, however, have resolved that the unauthorized removal of any tangible property, even though only temporary, and the copying of such property to extract a trade secret that is then sold or otherwise unauthorizedly disposed of will be enough to meet the Act's requirements of goods, wares, and merchandise.

The proposed recodification of the Federal Criminal Law would place the provisions of present section 2314 into a new provision, 18 U.S.C. 1732. The language of the new section is considerably changed from that of the present one and uses the generic term "property," but the commentary on the draft states that no intention exists to add to the current federal law boundaries of "property" as now conceived.[4] Until amendatory legislation or clear judicial decision, the question poses some uncertainty in the application of the federal statute.

The second problem is that of jurisdictional amount. The value must be at least $5,000 (so as not to clog federal criminal justice administration with insignificant matters). But the method of valuing is not spelled out, and different courts have taken different positions. Some have argued that market value is the only possible measure, and others that intrinsic worth is proper. A situation in which no market yet existed for a trade secret that could nevertheless be shown to have a reasonably calculated intrinsic worth of millions might result in dismissal of federal criminal prosecution against the thief.

To the extent that it is available and applicable, the federal Act should be considered among measures used to suppress industrial espionage. There is deterrent value in public knowledge that a given enterprise will resort to all available remedies, including criminal prosecution. If a solid federal case could be made against the identified industrial espionage agent for interstate transportation of stolen goods, then a good case might also be possible against his collaborators either for receiving the stolen secret (the following section of the law, 18 U.S.C. 2315) or for conspiring to commit the crime of interstate transportation (18 U.S.C. 371).

Because there is both personal and corporate criminal liability, combined successful prosecutions of the agent and his principals could have a profound effect not only on the persons involved but also on the firms using or intending to use the stolen trade secret. Of course, criminal conviction requires proof of guilt "beyond a reasonable doubt," a much heavier burden of proof than the "preponderance of the evidence" that is the burden in civil suits. However, an adequate factual situation, often made possible by the extent of the protective measures taken before the theft, and alacrity and vigor in preliminary investigation by the principal victim may permit an early criminal complaint and satisfactory conclusion. In later civil litigation, admissions made in the criminal trial are admissible. A criminal conviction may make a civil recovery easier, but an acquittal of criminal charges will not necessarily extinguish the civil action.

FEDERAL PATENT LAWS

There is an intimate relation between patents and trade secrets, although the two represent directly opposed concepts of invention management. The patent laws provide that an inventor who first develops a new machine, manufacturing process, composition of matter, plant, or design that is sufficiently novel and useful can apply for and receive an exclusive right to that invention for a period of 17 years.[5] The requirements for obtaining the patent, however, are quite specific. The invention not only must be novel and useful but must represent a positive contribution to the state of the art "beyond the skill of the routineer." That is a higher-order utility than merely requiring that information be useful before it can qualify as a trade secret. A much lower level of novelty or newness (although still something more than the obvious) will qualify information as a trade secret. For that reason alone, the choice must often be made to treat information as a trade secret rather than to seek a patent.

Beyond novelty, a patentable invention must be fully disclosed in its best mode so that a person reasonably skillful in the art required could apply it. Again that is the obverse of a trade secret, which in fact is lost by that very kind of unrestricted disclosure.

The exclusive right to patent protection expires after 17 years, and the world is then free to use the invention. The trade secret, however, can be applied indefinitely as long as it continues to meet the trade secret tests.

The rationale underlying patents is that, in return for the full disclosure of an invention whose use will benefit the common-wealth, the inventor is granted an exclusive right to the initial period of such use. Disclosure is the consideration for exclusivity. The rationale underlying trade secrets, on the other hand, is that one who develops a useful thing or process may enjoy the exclusive benefits of it only as long as it remains secret. Moreover, another who independently and without improper means develops the same secret is entitled to the same protection. And if an independent developer chooses to make public disclosure of his invention, that may effectively terminate the original developer's rights, because the secret will then have entered the public domain. Whereas patents grant exclusivity on condition of disclosure, trade secret law affords a qualified protection of exclusivity only on condition of secrecy.

There is no industrial espionage target in a patented invention (although there may well be one prior to the filing of the application), because any use not licensed by the patentee constitutes an infringement. Anyone can purchase a copy of an issued patent from the patent office and learn all that the inventor has disclosed. Of course, there are many cases in which a user infringes; that is, he makes, uses, or sells a patented invention without a license. In many ways the clandestine fashion in which he does so can resemble an espionage operation, but it is not our purpose in this limited work to discuss patent infringement beyond noting that the same techniques an espionage agent might use to steal a trade secret could be and sometimes are used by a patentee who is attempting to discover whether his patent is being infringed.

The difference in motivation, however, will not protect the patentee from civil or criminal sanctions if the methods he uses constitute crimes or torts. The techniques of the industrial espionage agent can be used to gather the required information. Therefore, a sound security program to protect trade secrets and other confidential information may have the additional value of deterring improper inquiries or inspection by those who suspect infringement of their patents.

Information regarding a patentable invention, prior to the time when a patent application is filed, must be protected with the same degree of care as a trade secret would be. A patent will not be issued if use or publication has occurred earlier than one year before the filing. A stolen trade secret that is reduced to use amounting to a disclosure that is not discovered by the original developer until more than a year later can no longer be patented. By then the disclosure may have been public or may have been made to innocent persons now also using the invention so as to destroy the original developer's rights against everyone but the thief. Cold comfort if the thief is not the real competitor!

While a patent application is pending, the information disclosed therein remains eligible for trade secret protection because the disclosure to the patent office is not deemed a public disclosure. It is the issue of the patent, not the filing of the application, that puts the disclosed information into public domain, subject for 17 years to the patentee's exclusive rights.

A recent Supreme Court decision [6] has created widespread concern over the state of the law regarding enforceability of agreements to license the use of trade secret information that thereafter is the subject of a patent application. It is presently unclear whether any agreement looking to royalty payments during the prepatent period of trade secrecy protection is valid. But it has been decided that the right to royalties for use of a patented invention is not perfected until the patent is issued. Thus a licensee who contests the validity of a patent can suspend payment of royalties until a judicial decision has been reached. If the patent is not sustained (a not infrequent result), no back royalties are payable. If, in future the law is held to be that an agreement to treat information constituting a patent application as a trade secret is not valid, as has been suggested in one recent trial court,[7] the owner of such an invention may lose all his rights.

Although the status of a trade secret embodied in a pending patent application is in some confusion, it has been settled that the trade secret, if relied upon exclusively in that form, will be protected by the common law principles previously discussed. The proprietor of confidential information that may have the status of a trade secret is still serving his own best interest by protecting that information to the maximum practicable extent. License arrangements, now more than ever, represent serious risks to the proprietor. It is not inconceivable that an unscrupulous competitor, in reliance on the unsettled law, may negotiate in apparent good faith for a license as a way to secure secret data while he privately intends to resist efforts to enforce the agreement.

STATE CIVIL LAWS

Although the substantive treatment of patents is reserved exclusively to the federal government, other areas of the statutory law are shared by federal and state jurisdictions. Unfair competition and antitrust statutes can be found in the several states for the regulation of purely intrastate commerce. For the protection of confidential information in such statutory intrastate contexts, the provisions of the local law must be consulted.

Generally, there will be close parallels between counterpart provisions in federal and state law regarding unfair competition and restraint of trade. In some situations there will be a choice of law and forum, and an aggrieved proprietor of confidential information will have to determine the appropriate selection on the basis of subsidiary issues such as statutes of limitations and procedural matters such as discovery. Those questions are beyond the purpose of this work; they are noted to indicate the range of research required in preparing a civil legal action or defense in matters involving confidential information.

STATE CRIMINAL LAWS

Among the best weapons in a trade secret proprietor's defensive arsenal are the state criminal laws. The states listed in the accompanying box have developed specific theft statutes that identify, define, and establish trade secrets, as such, as the stolen property.

STATE CRIMINAL STATUTES
COVERING THEFT OF TRADE SECRETS

Arkansas	Ark. Stat. 41-3949
California	Cal. Ann. Penal Code 499c
Colorado	Col. Rev. Statutes 40-5-33
Georgia	Geo. Crim. Code 26-1809
Illinois	Annot. Stat. 38-15
Indiana	Statutes 10-3048
Maine	Rev. Annot. Stat. 17-2113
Massachusetts	Annot. Laws 266-30
Michigan	Comp. Laws Annot. 39-752.771
Minnesota	Statutes 40-609.52
Nebraska	Rev. Stat. 28-548.01
New Hampshire	Rev. Stat. Annot. 580-32
New Jersey	N.J. Annot. Stat. 2A:119-5.3
New Mexico	Statutes 40A-16-23
New York	Penal Law 155.00(6), 155.30, 165.07
Ohio	Rev. Code Annot. 13-1333.51
Oklahoma	21 Ok. Stat. Annot. 1732
Pennsylvania	Statutes 18-4899.2
Tennessee	Title 39 Tenn. Code Annot. Sections 39-4238 *et seq.*
Wisconsin	Wis. Stat. Annot. Criminal Code 943.205

In those states there are varying approaches to the problem, beginning with varying definitions of what constitutes a trade secret. The New Jersey statute, for example, defines a trade secret as "the whole or any portion or phase of any scientific or technical information, design, process, procedure, formula or improvement which is secret and of value." Thus the problem of whether the information itself or only the medium for conveying it constitutes the secret would appear to be resolved in favor of the information.

In the New York statute, "secret, scientific material" is made the subject of theft and is defined as "a sample, culture, micro-organism, specimen, record, recording, document, drawing or any other article, material, device or substance which constitutes, represents, evidences, reflects, or records a scientific or technical process, invention or formula, or any part thereof." Thus the same problem is posed in New York as in the Federal Stolen Property Act, namely, whether the theft of a pure idea is criminal.

Moreover, the terms "scientific" and "technical" could be argued not to include other business data such as costs, methods, and corporate opportunity. And New York arbitrarily assigns thefts of "secret, scientific material" to the rank of third-degree larceny and makes them punishable by a maximum of four years' imprisonment. Although secret scientific material might be proved to have a value far in excess of it, $250 is the value limit in New York on third-degree larceny. Another section of the New York statute makes theft of property over $1,500 in value a second-degree larceny with a maximum sentence of seven years, but the specific mention of "secret, scientific material" in the third-degree larceny section and its exclusion from the second-degree larceny section probably means that the higher penalty could not be sought even if the property stolen were shown to be worth millions.

A complementary section of the New York Statute (165.07) settles the copying-without-removing difficulty and makes it a felony (four-year maximum sentence) to make a tangible reproduction or representation of "secret, scientific material" with an intent on the part of the copier to appropriate the use of such material to himself or another without right or reasonable grounds to believe there is a right to such material. Again, the actual value of the secret material does not control.

States that have not enacted specific trade secret penal laws will nonetheless have larceny statutes of a more conventional type. Criminal prosecutions can be developed under them, with due regard to the recurring difficulties of proving a trade secret to be property and of establishing its value. However, increased interest and attention are being paid by legislators to the problems of trade secret and confidential information protection under state criminal law.

To the extent laws are available, they should be used. If present laws are inadequate, proprietors of confidential information and trade secrets should support efforts to improve them. If a person bent on industrial espionage faces only the civil penalty of inability to use the stolen secret or the requirement to make the victim financially whole, his potential gain, if successful, may still be attractive in enough cases to encourage continued spying. If criminal penalties consistent with the real nature of the crime—the theft of someone else's secret—are available, the prospect of imprisonment and all the other undesirable consequences of a criminal conviction may be a more effective deterrent.

REFERENCES

1. See bibliography in R. Milgrim, *Business Organization—Trade Secrets,* Rev. Ed. (Albany, N.Y.: Matthew Bender & Co., Inc., 1972).
2. *Restatement of the Law of Torts,* 1939, Section 757, Comment b., Committee on Torts, American Law Institute.
3. *B. F. Goodrich* v. *Wohlgemuth,* 192 N.E. 2d 99, 117 Ohio Appeals 493.
4. *Working Papers of the National Commission on Reform of Federal Criminal Laws,* Vol. II (Washington, D.C.: G.P.O., July 1970), p. 916, par. 4.
5. 35 U.S.C. 1 et seq.
6. *Lear* v. *Adkins,* 395 U.S. 653.
7. So held by the District Court (S.D.N.Y.) in *Painton* v. *Bourne,* 309 F. Supp. 271, but reversed as to that finding by the Circuit Court of Appeals, 2d Circuit, in *Painton* v. *Bourne,* 442 F. 2 216.

4

HOW TRADE SECRETS
ARE LOST—
THE INTERNAL THREAT

A television or movie story about industrial espionage would probably have a rather sinister character, the industrial spy, using bugging devices to listen in on the victim's conversations either in an
office or by telephone. If comedy were intended, the spy might be
portrayed as awkward and not very bright. He would probably
obtain the wanted information in spite of rather than because of
his clumsy efforts, and it would doubtlessly be through the amateurish use of electronic listening equipment.

By exposure to such fictional situations, many business executives
may have been led to believe, first, that industrial espionage is not
a threat because only clumsy characters are involved and, second,
that his business secrets will be safe if he protects them against
electronic eavesdropping.[1] Both beliefs would be wrong. Few real-
life spies are as inept and clumsy as the "Mr. Crest" of Chapter 1,
and electronic bugging, although a definite threat, is far from being
the only way in which valuable business information may be lost.
This chapter and the one following are devoted to a discussion of
the variety of ways in which an enterprise may be victimized.

In general, however, an organization may lose valuable information in either of two ways: as a result of the activities of an individual or individuals working inside the enterprise or as a result of activities of an individual or individuals operating from the outside. And, of course, losses may also result from the combination of the two. This chapter deals with the internal threat, and Chapter 5 deals with the external one.

The Employee

Probably the most serious internal threat to trade secrets is the employee. When an effort is made to determine how information might be lost, it must be remembered that every employee in the organization may be in a position to disclose proprietary data. Employees who have sensitive information in their possession or who have access to it, such as those engaged in research and development, planning, and top level decision making, must obviously be included in an industrial espionage prevention program. But there are other potentially dangerous employees who may be overlooked because their handling of trade secret information is less obvious. For that reason, the vulnerability of some of those employees will be discussed later in this section.

Release of sensitive company information may be due to carelessness or lack of knowledge, or it may be an intentional act in response to a bribe or other motivation. The results are equally serious; that an employee did not intend to give out information is of small consolation if the enterprise has been damaged. Therefore, any industrial espionage preventive program must be designed to protect against both unintentional and intentional disclosures of information by employees.

THE DISLOYAL EMPLOYEE

Top management representatives in most organizations believe their key employees and executives are loyal and usually make every effort to insure that they remain so. However, it is not always possible to satisfy everyone, and a number of recorded cases of industrial espionage indicate that disloyalty is often manifested by a disgruntled employee. The individual probably begins work with the company as a loyal employee and would not think of intentionally taking any action that might injure the organization. However, because of changes in his outlook or because of events taking place in the organization, his attitude may undergo a radical change over a period of months or years. Any number of factors might influence the change: an emotional problem, sexual difficulties, drugs, alcohol, gambling, excessive indebtedness, or family troubles.

The constant increase of prices and living costs in recent years might also influence the loyalty of an employee, especially an executive or key employee, who is frustrated because his compensation does not keep pace with tax increases and the price spiral. Any employee may feel that he should rise faster in the organization and earn more money. Lack of recognition, as well as failure to receive promotion or salary increases, may contribute to the dissatisfaction of an individual. Because of frustration and resentment he may take the opportunity to engage in industrial espionage by pirating valuable company secrets that may be readily available to him.

The mergers and reorganizations that have taken place in many companies in recent years have contributed to employee disaffection. Changes in personnel and other general changes may cause employees to become disloyal through concern for their future financial security. Any such employee might steal company information in the hope of using it to obtain a position elsewhere. Alternatively, he might plan simply to sell the information to another company or to use the data in a business he himself would establish in competition with his former employer.

One disgruntled employee engaged in large-scale thefts of secret technical data and cultures used to produce antibiotic drugs from American Cyanamid's Lederle Laboratories.[2] The case also demonstrates how an employee can utilize his employer's sensitive information to set up his own business. Dr. Sidney Fox, during his five-year term of employment at Cyanamid as a chemist, increasingly felt that his work was not appreciated and that he was underpaid. He began stealing secret data and cultures that had been developed at great expense. He sold the material to Italian competitors of American Cyanamid.

Initially, Fox made his sales through intermediaries, but later he began to deal directly with the Italian companies through a cover company he called Kim Laboratories. He operated Kim Laboratories as an unincorporated business a few miles from the Cyanamid facility where he worked. He later resigned from Cyanamid, reorganized Kim Laboratories as a New York corporation, and became president of the company. When he left the Cyanamid facility, he took with him a supply of proprietary data along with antibiotic and steroid starting materials. He also arranged with a fellow employee at Cyanamid to continue bringing him data, and he made the employee an officer in Kim Laboratories.

Cyanamid sued Fox for damages and was granted the relief sought. The court held that Fox had come into possession of legitimate trade secrets by unlawful means. Fox was also prosecuted and sentenced in federal court on the charge of interstate transportation of stolen property.

Another manifestation of disloyalty, although it might appear to be otherwise, is the leaking of vital company information to financial institutions or the press by executives or key employees. The individual may be motivated by a desire to feel important and to demonstrate that he is intimately familiar with the operations of his organization. Also, an employee might intentionally disclose information to the news media or a competitor, without expectation of financial reward, because he believes that others have a right to the information. An example of such an action was the release of the Pentagon papers to the press by Daniel Ellsberg in June 1971. In that case government information rather than business or industrial secrets was involved, but an individual with the same ethical motivation as Ellsberg might conceivably release proprietary information because he felt strongly that it should be shared with others.

THE MOONLIGHTING EMPLOYEE

An employee may decide to use his free time either in working part time for a competitor or to set up his own business. He may be motivated by financial pressure, or he may want to progress faster in his chosen field. In any event, the moonlighting employee might steal company secrets from his regular employer for use in connection with his outside activity, or he might use his employer's secrets to develop a new product to compete with his employer's product.

Moonlighting is not restricted to any class of employees. Because it may take a multitude of forms, it may be practiced by anyone from the highest-paid executive to the lowest-paid hourly worker. It has become a growing practice, and it will probably continue to grow as employees gain more free time by the reduction of hours in the workweek. It is interesting to note that contract guard agencies commonly employ individuals who are working regularly elsewhere and moonlight as guards to obtain extra income.

In most organizations the attitude toward moonlighting is based on whether there is a conflict of interest or whether the outside employment activities of an employee would cause damage to the organization. Some companies discourage moonlighting entirely because the strain of holding a second job or operating a business may reflect on the employee's efficiency and effectiveness in his full-time position.

THE MOBILE EMPLOYEE

In recent years, the ability of individuals with unique skills to move from company to company has created a special trade secret hazard. That has been especially true of research workers, engineers, and scientists because of the explosive developments in science and technology. Supporting personnel such as draftsmen, technicians, craftsmen, clerical assistants, and maintenance workers have also become mobile employees and so have contributed to the problem of safeguarding proprietary information.

Top management representatives of any organization would probably agree with a number of court decisions that individuals should not be prevented from utilizing their naturally acquired skill and experience when they join a new employer. On the other hand, they would probably also agree that the mobile employee must be prevented from stealing proprietary information for the benefit of his new employer and to the detriment of his former employer. Because a number of legal questions are involved, the problem of the mobile employee will be discussed in detail in Chapter 10.

A 1967 case, tried in the New York supreme court, illustrates the problem.[3] Peter Schenk and Michael Lauro were engineers at Republic Aviation, where, for a period of 18 months, they played key roles in the design and development of a complex integrated aircraft weapons monitor and control system designated Comac. The two engineers left Republic Aviation in January 1965 to work for the Technical Measurement Corporation. Within a few months it was learned that their new employer had developed a weapons system very much like Comac. What was of interest to Republic was that Technical Measurement Corporation had never been in the weapons systems business before it employed Schenk and Lauro. The suit filed by Republic against Schenk and Lauro and their new employer alleged that the Comac system had been pirated by the two engineers. The court agreed.

THE MARKETING EMPLOYEE

Salesmen and other marketing personnel are constantly in the position of having to convince outsiders that they represent outstanding organizations and that their products are equal to or better than those offered by competitors. Also, they are often extroverts who are interested not only in enhancing the product and company image but demonstrating how important they are as individuals and how much they know. Because they are usually in daily contact with a wide range of individuals, including competitors' representatives, they have a great opportunity to disclose sensitive company information that might be used to do enormous harm to their employers.

Marketing personnel may not only be entrusted with confidential information so that they can effectively sell the company product but also be in possession of customer lists or have valuable information about clients. Such information may be very attractive to competing organizations. On the other hand, because such employees can often be expected to be creative, their access to sensitive product information should not be limited—many of them can make recommendations for changes and improvements in products. Therefore, although any industrial espionage prevention program must include them, marketing employees should be allowed adequate information so they can perform effectively.

THE PURCHASING EMPLOYEE

Procurement employees should have a knowledge of new developments in the organization so they can obtain the best available material at the most favorable prices. Also, a supplier can often make significant contributions to a new development if he knows what is going on. However, giving too much information to a supplier can be dangerous; a buyer may gratuitously transmit to a vendor's representative information that is of great value. If the supplier happens also to be selling to a competitor, he may pass the information on in an effort to secure a favored position with the competing company.

Vendors, of course, are interested in influencing the decisions of buyers, and they have been known to resort to bribes in the form of cash or gifts. If the vendor representative finds a buyer tempted by such offers, he may use the same technique to solicit sensitive proprietary information that he can pass on to a competitor. Like marketing personnel, procurement employees must have sufficient information to perform effectively. However, the possibility of the release of sensitive information by this class of employee must be recognized when an industrial espionage prevention program is developed.

THE CONSULTANT

Although consultants may not be regarded as employees in many organizations, they are usually given the free access to company facilities and information that is accorded to employees. That is necessary because if the consultant did not have free access to company data in the areas in which he has an interest, his reports would be of limited value to the enterprise employing him.

Consultants are now available in every area of the administrative side of business, including security, marketing, production, research, and technology. And because consultants can often contribute to the success of an organization by virtue of their objectivity, independence, and wide experience, they are frequently utilized. Over a period of time, the same consultants may be engaged by several competing companies. Usually they have high ethical standards and would not intentionally betray confidences or engage in any activities related to industrial espionage. However, the same hazards of release of sensitive information, either intentionally or unintentionally, arise with consultants as they do with employees. For that reason, a program designed to protect the enterprise against industrial espionage activities of employees must include consultants as well.

One example of the use of information received by a consultant from one company and used in the development of a competing product by another company involved Rise shave cream.[4] Norman Fine was an employee of Foster Snell, a consulting firm that was retained by Carter Products. Fine, while assigned to the company, was a key figure in the development of Rise shave cream.

Later, Fine was hired by Colgate-Palmolive to assist in the development of a different pressure-propelled shave cream. Soon after the new Colgate product was marketed, Carter Products obtained an injunction against its sale on the ground that Fine had apparently applied secret technical information he had developed at Carter Products. After 13 years of litigation, it was decided by the court that Fine could not possibly have done the job he was assigned at Colgate-Palmolive without utilizing the trade secrets he developed at Carter Products. For that reason, Carter Products was awarded damages.

CONTRACTOR AND VENDOR EMPLOYEES

Although they are not employees of the organization, contractor and vendor representatives may be discussed with employees in regard to access to the facility. They might also be included in the next chapter, which deals with the external threat, because they actually represent outside organizations. However, because of their frequent close working relationships with company employees and because they perform their work inside the organization, the hazard they represent is included here.

Contractor and vendor employees may represent a wide range of skills from the lowest-paid hourly worker to the highly skilled professional. They include janitors, maintenance workers, guards, vending machine and food service workers, telephone company servicemen, and computer service personnel and also secretaries, clerks, technicians, and engineers whose services are contracted for during heavy workload or when the organization, for some reason, does not want to hire a permanent employee.

In considering a worker of that type, it must be remembered that he may be working in the facility when regular employees are not present and that he will probably have little or no supervision from anyone who represents the facility. A member of a contract janitorial force is a good example. Such an employee must generally have access to all areas, and he normally performs his work after the close of business. A contract janitor usually works alone, and he has ample time to collect information if that is his objective.

The contractor or vendor may neither encourage nor condone industrial espionage by its employees, but many of those employees are paid low wages and have no motivation to be loyal to either their own employer or the company to which they are assigned. As a result, one of them might be easily bribed with a small amount of money.

MISCELLANEOUS CATEGORIES OF EMPLOYEES

A variety of other employees who have access to sensitive material should be mentioned. The secretary, for example, may be overlooked when an industrial espionage prevention program is being designed, even though she may have a large amount of proprietary data at her disposal. She may be privy to all of the sensitive data in her supervisor's possession, and she will usually have access to all of the written material and reports available to him. In the normal course of her work she might unknowingly disclose secret information to competitors and others. For example, a remark on the telephone to a competitor's representative that might appear to be completely innocent to the secretary might give the competitor exactly the information he is seeking. Her explanation that her boss is unavailable because he is away from the office for a week in a certain town or country might be of great benefit to a caller who already suspects that a merger negotiation or some other significant arrangement has been pending with a company at that location.

Service employees, such as motion-picture operators and workers in the executive dining room and similar locations where sensitive company information may be discussed, must also be considered. Janitors, maintenance workers, and other service personnel who have free access at any time to the offices of the top executives or sensitive research areas must not be overlooked.

Electronic data processing and computer center employees must be given particularly careful consideration when the industrial espionage prevention program is set up. The hazard is merely mentioned here because a detailed discussion is included in Chapter 8.

Publications

Information that is published by most organizations can be divided into two general classes: (1) material that is used within the organization as in a company newspaper and (2) material released for consumption by outsiders in such form as sales brochures. Because the courts have ruled that, once its unrestricted publication has occurred, information is considered to be in the public domain and can no longer be regarded as secret, both classes of publication are a potential hazard to the protection of proprietary information.

An organization naturally wants to put its best foot forward in the release of information, and every effort is usually made to create a good impression. It may happen that so much effort is made to stress the importance of the enterprise and the work being done that sensitive company data are revealed. It can safely be assumed that competitors will review information released within the organization as carefully as they do that released outside it for data that may reveal company secrets or offer clues that will give them an advantage.

A press release can be carefully reviewed to insure that no secret data are included, but a press conference or interview held in connection with the release or following it can constitute a special hazard if it is not carefully planned and controlled. Experienced interviewers will make every effort to obtain as much information from it as possible. Once information that should not be released is revealed in such a situation, the results can be devastating, because it will usually be impossible to prevail on the news media to not release it.

Another publication hazard that must be considered is the preparation of papers and articles by employees for release in professional journals or other publications. Generally, in years past, that problem involved only scientists, engineers, researchers, and teachers, who were guided by the publish-or-perish rule. Now, because of the proliferation of society and professional publications in all fields, the problem of information release is related to all areas of specialization.

The technical secrets of the switching systems used by telephone companies are reported to have been released in a technical journal. The result was the development of beeper boxes that allow individuals popularly designated as phone-phreaks to circumvent the toll system and make calls anywhere in the world without paying for them. The release of the data had a serious impact on telephone companies not only because of the telephone toll revenue being lost but because it is reportedly possible for the telephone system in an entire city or a large area to be completely tied up by a knowledgeable phone-phreak.

Employees who are taking courses in universities, colleges, or technical institutes may desire to publish papers or theses based on knowledge they have obtained while working at their jobs. Such education efforts are useful to the employee and the organization and are usually encouraged by management, but there is nonetheless a danger that trade secrets might be unwittingly incorporated into the submitted material.

The publication for limited internal distribution of sensitive reports that contain research data, marketing studies, financial data, or other proprietary information is usually tightly controlled. However, a program designed to protect the sensitive data should include the employees responsible for the mechanical aspects of the publication, such as writing, editing, drafting, and reproduction. The best-designed program to insure protection of the information once it is published will be of little value if those employed in the publication process are given an opportunity to obtain copies of the material while it is being published. Another hazard that is related to the publication of such reports is the disposal of scrap. Rough-draft and other scrap material can be just as valuable to an industrial spy or competitor as the final report.

Seminars, Conventions, and Trade Shows

If an organization has arranged for an exhibit at a meeting, it will normally make every effort to project the most favorable company and product image. The techniques utilized may include elaborate displays, working models, color movies and slides, attractive girls, company brochures and handouts, and a representative to answer questions. As is true of all company handouts, the display and the material available for distribution must be carefully screened to insure that no sensitive data are included. Those in attendance at the exhibit must be cautioned about releasing information during discussions with people who indicate an interest in the material being shown.

A meeting is usually held for the very purpose of exchanging information. The organizational representative who attends the meeting hopes to obtain useful data in his field of specialization, and he may give information to other attendees if he is scheduled to present a paper. However, it is not necessary that the employee make a formal presentation at the meeting to have him disclose information; he will have ample opportunity to discuss affairs of his enterprise during the question and discussion periods that are usually a part of the organized program. In addition, he will have an opportunity to enter into discussions with other attendees at coffee breaks and whenever else the formal sessions are in recess.

An area that can be particularly hazardous at a meeting is the hospitality suite. A pleasant atmosphere, friendly associates, and a bar dispensing the best brands of liquor in unlimited quantities comprise the usual setting. A normally discreet employee may be influenced by such an atmosphere to throw caution to the winds, develop a loose tongue, and discuss proprietary data with a competitor's representative or an industrial spy who is there precisely because he knows it is a good place to collect information.

REFERENCES

1. Shepherd Mead, *How to Succeed at Business Spying by Trying* (New York: Simon & Schuster, 1968), a novel about industrial espionage.
2. *American Cyanamid* v. *Fox,* 140 U.S.P.Q. 199.
3. *Republic Aviation* v. *Schenk,* 152 U.S.P.Q. 830, 834-835 (1967).
4. *Carter Products, Inc.* v. *Colgate-Palmolive Company,* 104 U.S.P.Q. 314, 108 U.S.P.Q. 383 (1956).

5

HOW TRADE SECRETS
ARE LOST—
THE EXTERNAL THREAT

Anyone who must face the hazard of industrial espionage will probably think first of the spy, the most serious external threat. There are, however, a number of other external threats that should not be overlooked. They are discussed in this chapter.

The Industrial Spy

Industrial spies have been classified into two categories by some writers—the professionals and the amateurs. Both are equally dangerous and are capable of doing great harm to an organization.

The effective spy will rarely steal a document or other material that the owner can identify as being missing. Instead, he will memorize the information, copy it, or photograph it. He is a patient, meticulous worker; and if it is necessary, he will devote months to the task of obtaining the information he wants. Usually the spy will look for the easiest way to his goal and will exert no more effort than is absolutely necessary. He will use any method and will not be hampered by legal, ethical, moral, or financial considerations. In addition, he will take maximum advantage of human frailties.

The amateur industrial spy may be an opportunist who has happened onto material or information that he recognizes would be of value to an individual or organization. He also may have been hired by someone in the same way a professional would be hired to obtain certain identified data. Because of a lack of experience and professional knowledge, he may not be as skillful as the professional. He will probably be frightened and in a hurry to capitalize on his efforts. However, the amateur may use many of the techniques used by the professional, and he may be equally successful in the collection of material. The results of the activities of either the professional or amateur can be disastrous to the victimized organization.

In recent years the activities of the industrial spy have taken on an international character. Not only are trade secrets marketed in foreign countries by industrial spies who may be U.S. citizens but, contrary to a naive belief held by some, the spies who represent foreign powers do not limit their activities to military secrets. The late J. Edgar Hoover in his writings and comments constantly stressed that the Soviet Union is interested in obtaining every kind of industrial data. Japan is reported to have established a training school for industrial spies in Tokyo.[1] The all-encompassing activities and the intrigue of a ring organized by the East Germans to obtain industrial secrets was described in *Newsweek* in 1965:[2]

Envious of the booming abundance of Western Europe, the fumbling East German regime of Walter Ulbricht in the late 1950s decided to narrow the prosperity gap with a melodramatic program of industrial espionage. Using the code name Operation Air Bubble, the government deployed a network of agents across the continent to pirate technical information that would cut corners on research and reduce production costs in inefficient industries. But Ulbricht's Air Bubble was decisively punctured last week when a French State Security Court convicted a young East German hydraulics engineer of industrial spying.

The engineer, fair-haired, 32-year-old Herbert Steinbrecher, was arrested in the Pigalle section of Paris last fall carrying microfilm disguised as candy. He had been on the way to southwestern France to establish a consulting engineer's office as a front for carrying out Air Bubble's audacious "master plan"—stealing the secret plans for the British-French Concorde supersonic airliner from Sud-Aviation in Toulouse.

A sequel to that story appeared in the same publication in 1971: a feature article describing the expulsion of 105 Soviet officials from Britain because of their espionage activities. According to the article, the Soviet spy effort was being concentrated on industrial espionage to help close the technology gap between the Soviet Union and more advanced nations. The results of Operation Air Bubble as they relate to the theft of the Concorde secrets mentioned in the 1965 article are described in the 1971 article as follows: [3]

> Sometimes, the victim never even knows he's been had, but sometimes, too, the fruits of industrial espionage are spectacularly conspicuous. The obvious similarities between the Anglo-French supersonic transport, the Concorde, and the Soviet SST, the TU-144, are more than coincidental. (In fact, the outraged British have dubbed the Russian aircraft the "Concordeski.") Many of the Concorde plans were reportedly stolen by East German master spy Herbert Steinbrecher. Converted to microfilm, the plans were stuffed into a tube of toothpaste, which was dropped off for a Soviet courier in a toilet on the Ostend-to-Warsaw express train. Possibly because of this exploit and other Communist coups, the TU-144 made it into the air several months before the Concorde.

An individual interested in hiring an industrial spy would get nowhere by searching the telephone directory yellow pages or business directories under such headings as "industrial spies," "spies," or "industrial espionage." That should not lead anyone to conclude that industrial spies do not exist.

Many thousands of agents were well trained in the techniques of obtaining information of all types by the various military services and by other government agencies over the years during and since World War II. The training often included both special schooling and practice of the art under highly qualified superiors. Today a number of those spies are applying their experience and training in U.S. business and industry. Because they operate clandestinely, there is no way to know how many of them there are.

Some individuals with espionage training who want to specialize in obtaining business and industrial secrets establish investigative or detective agencies; others prefer more nondescript designations for their activities. For example, a one-man operator has adopted Inter-American Security Consultants as his firm name.[4] Since the purpose of this book is to deal with the prevention of industrial espionage and not to recommend the use of industrial spies or to tell the reader how to locate one, no further effort will be made to establish that spies do in fact exist. The remainder of this section is devoted to a discussion of the techniques the spy can be expected to use, so that protective measures can be planned.

THE PATSY

The method regarded by most operators as the best one is to use a patsy—someone who works for the target organization and who will cooperate willingly or under duress. The patsy may be any one of the employees mentioned in the preceding chapter. The spy works with the patsy over some period of time, and initially he does not ask for the type of information he really wants. He gradually asks for more data, but he never lets the patsy know which information is important. At some time in the negotiation, the spy hopes to obtain the information he wants without letting the patsy know how valuable it is. Nor does the spy usually reveal the identity of his client to the patsy. Even if the conspiracy becomes known, the client is not compromised.

The patsy may be motivated to cooperate by any number of factors. A bribe is often successful, and that technique is usually preferred because it is regarded as simpler than any other. Locating an employee who is in financial difficulty is usually the first step in developing a candidate who can be bribed. One professional spy described how he located employees vulnerable to bribery.[5] He recorded the license plate numbers of all the cars in a parking lot adjacent to the facility from which the information he wanted could be obtained. He then got the names of the car owners and checked them against credit bureau files. Of about seventy-five persons he found three to be in serious financial difficulty. They were his prime candidates.

The corruption of the employee usually starts slowly. At first, only information that can be easily obtained is requested and payment is made in cash. Later, the spy asks for data that are more difficult to obtain and pays by checks so he has the victim's endorsements as a record of the bribes. Then if the employee refuses to supply needed information or is otherwise uncooperative, the spy threatens to expose him and uses the endorsed checks as a form of blackmail.

If the patsy is a man and a bribe would not be effective, the industrial spy can always consider using an attractive female. She might be able to obtain the information needed over a period of time; but if she fails, the spy can always resort to blackmail as a means of dealing with the individual directly. If the patsy is a female, the spy himself or one of his associates may use romance as a means of obtaining data. One professional spy was quoted as saying that an attempt should never be made to bribe a secretary because she usually has a feeling for her boss that borders on love.[6] He advised that both sex and romance should be used and that a secretary should never be asked for information directly in violation of her loyalty to her supervisor. Instead, information can be obtained in conversation, and the employee will be unaware that she has betrayed her boss and her organization.

The industrial spy can be expected to attempt to locate an individual in the organization who has a grudge against either an executive or the entire company. Such a malcontent may offer information for no other reason than to get even and satisfy his bruised ego.

THE UNDERCOVER OPERATOR

If the industrial spy is unable to develop a source of information within the organization, he may place his own representative in the enterprise as an employee. He may have an associate or an employee he can use, or he may hire an individual to go to work in the organization. The following is a description of the undercover methods used by one operator: [7]

A Washington, D.C., private detective, specializing in undercover operations in the small experimental electronics companies of the surrounding area, usually begins such an assignment by putting a want ad in the paper for a suitable job classification. Technicians, engineers, electricians, and maintenance personnel are preferred to desk men because of their ability to move freely about without attracting suspicion.

A satisfactory applicant is selected by personal interview from those responding to the ad; then the pitch begins: a certain type of "general" information is needed from the "Electroray Company"; if the applicant can land a job with them (a relatively easy task in this high-turnover industry) the detective will supplement his income by as much as $60 a week, with bonuses for especially valuable tidbits. With the prospective spy going on the detective's payroll immediately, awaiting employment, indoctrination begins: "I tell him how to write reports, what to look for, how to photograph documents, or get them out at night and bring them back in the morning. If the man is a technician, I interest him in being a detective someday and I pretend something quite different from what I'm actually trying to do. You gradually work him into what you are doing and tell him as little as possible. He doesn't know who your client is, or why you want the information, or which information is really important. That way, if he gets caught or outlives his usefulness you have no trouble. On cases where there's only a limited amount of money available, I advertise for a man to do part-time work on Sunday or in the evenings. Sometimes in the answers I find a man actually employed in the bank or business I want to get into. This is the guy I'll hire, give him some innocuous make-work until I can win his confidence enough to get the particular information my client is after."

TRESPASS

An industrial espionage agent may elect to enter the premises of the victim organization to obtain material he needs. He may adopt that method because he has not been able to use an undercover operator or develop a patsy within the organization or simply because the information he needs is so important or technical that he may prefer to do the job himself. There are also spies who prefer that method of operation.

A variety of methods may be used to gain entrance to the facility. The spy may pose as a reporter or as a government or municipal inspector or use one of any number of pretexts. Getting into the facility and obtaining the information he wants is often surprisingly easy. Once he is inside, if he is clever, he may be able to motivate employees to help him obtain information by taking him on a tour, demonstrating a process or activity, or allowing him to steal valuable data while no one is observing him. The methods used by one operator were described as follows: [8]

That job involves ———'s real specialty: the art of getting into a manufacturing plant without invitation and without arousing suspicion—a highly specialized craft at which ——— is a past master. He has by his own admission made his way into factories as a kindly old stockholder curious to know more about a process his company is using, as the overindulgent father of a hobbyist son (whose hobby conveniently varies from auto racing to boating to chemistry, according to the needs of ———'s client), as a fire or building inspector dutifully making his round, or even as a workman.

Once, when ——— was finding a particular plant hard to crack, he brazenly got into line with the morning shift, grabbed someone else's timecard and punched in, saw what he wanted to see and punched out again. On another occasion, he somehow enlisted the help of the police, who obligingly drove him to a plant on Long Island in their patrol car. There he excitedly pointed out to the night watchman a suspicious looking light on the roof. As police and watchman hurried up to investigate, he had no trouble at all locating the machine he wanted to inspect. When the frustrated searchers came down again, off went satisfied spy and police in the squad car.

For all the projects he takes on (he once testified in court that he had made 27,853 investigations for his many clients), ———— has no staff of apprentice and journeymen spies working for him. His is strictly a one-man operation. On the relatively few occasions when he wants to supplement his own investigative skills, he simply calls on his wife for help. In the role of a tirelessly curious magazine writer, she has scored some notable coups of her own. Once, for example, ———— dispatched her, after a thorough technological briefing, to the depths of an Oklahoma zinc mine to find out exactly what kind of device the company was using to dispel poisonous fumes from the diesel engines in the working area. And back she came, an hour or so later—with the story for her nonexistent magazine, and the facts her husband needed.

It can be seen from the preceding description of the techniques of penetration used by one professional operator that the ways to enter a facility are limited only by the imagination and ingenuity of the spy. Therefore, trespass is a serious hazard that must be guarded against in the planning of an industrial espionage prevention program.

LISTENING

The use of electronic listening devices to obtain proprietary data has for years been given a great deal of attention in the news media. However, in spite of the publicity given to the listening device disguised as an olive in a martini, a number of experts, including industrial spies, have been quoted as saying that the publicity given to that aspect of industrial espionage has magnified it far out of proportion to its actual importance. Mostly because of the attention that has been given to it, the subject is discussed in detail in Chapter 6. In addition to electronic bugging there are other means of listening for information that must also be included in a discussion of the methods used by the industrial spy.

Employees of most organizations have favorite meeting places such as bars, restaurants, and clubs. A favorite technique of the industrial spy is to mingle with employees when they are relaxing there during nonworking hours. The spy knows that the affairs of the organization will usually be discussed, and he is also aware that proprietary information may be included. As a result, he may only have to be a careful listener at the right time to obtain all the information he wants. He may also enter into conversations with employees whom he knows may have data in which he is interested. By guiding the conversation and asking adroit questions, the clever operator may be able to motivate individuals to give him information he could not otherwise get.

Obtaining information by using the telephone is another technique that may be used. A common practice is for the industrial spy to call an individual in the victim organization whom he knows has the information he needs and, by using some pretext, solicit the data. The spy may pose as a reporter, a potential customer, a fellow employee, or anyone he feels will be acceptable to the victim.

The spy may also contact the potential source of information in person and, while posing as a reporter, ask for information. By listening carefully, he may gain a large amount of the material he needs.

OBSERVATION

An industrial spy may develop information of value by observing the activity carried on in a facility or by watching individuals. He may do so by personal surveillance or by use of motion-picture or still cameras. Following top executives or key employees such as research personnel, design engineers, and marketing personnel to determine their contacts or activities is another technique of spying. The use of an airplane to obtain photographs of a facility was reported in *The Wall Street Journal* as follows: [9]

NEW ORLEANS—A Federal judge here has ruled in favor of Du Pont Company in what may become a landmark case in the annals of industrial espionage.

Judge Irving Goldberg of the U.S. Circuit Court of Appeals here ordered a Texas photography concern to divulge in a lower court the name of the party to which it sold aerial pictures of a Du Pont plant that uses a secret process to produce methanol. The chemical is used to make antifreeze and industrial plastics.

In March 1969 construction workers at Du Pont's new multimillion dollar plant in Beaumont noticed a low-flying plane circling over the still unfinished construction site and acting "in a generally suspicious manner," according to a Du Pont spokesman. After some detective work of its own, the company discovered that the plane contained a photographer who had snapped 16 pictures of the plant. The pictures would enable competitors to duplicate the Du Pont process, Du Pont alleged.

The photography company involved in the alleged aerial spying, Rolfe & Garry Christopher, declined to identify who had hired them. Du Pont brought suit in Federal district court that also asked for damages and an injunction to prevent further circulation of the photos and additional photographing of the plant. The court ruled in Du Pont's favor and the photography firm appealed.

In upholding the decision, Judge Goldberg noted that aerial photography was an unusual form of industrial espionage that didn't involve fraud or other direct violations of the law. "However," he said, "our devotion to freewheeling industrial competition must not force us into accepting the law of the jungle as the standard of morality expected in our commercial relations."

TRASH AND SCRAP

A way to gather information that has been a favorite of both U.S. and foreign intelligence agencies for years is to examine trash and scrap. The industrial spy also uses the method because he can expect to find a gold mine of information somewhere in the wastepaper and trash. In some organizations it is common practice for secretaries and others to dispose of rough drafts, spoiled copies, notes, and other such material in wastebaskets. The spy may be able to collect discarded material from the outside disposal area, or he may prefer to bribe a janitor or charwoman to save waste material from certain selected areas for him.

Models, mock-ups, dies, prototypes, and similar items may be as valuable as scrap paper. In Detroit, where at least $1 billion is invested each year in tooling and design changes on new car models, particular attention is given to protecting such material. As a counterintelligence technique, some companies are reported to have prepared elaborate decoy models to confuse industrial spies.

As an example of the popularity of the trash and scrap approach, an industrial spy has reported that on one occasion when he attempted to bribe a janitor to turn over scrap to him he found that he was too late. Another spy had already made an arrangement with the janitor for the material.

THE SURVEY OR QUESTIONNAIRE

The questionnaire has become popular in recent years as a means of collecting information that can be used to improve the competitive position of the organization making the survey and that may also benefit the individual who receives the questionnaire. The industrial spy can be expected to use the questionnaire in his effort to get proprietary information.

The spy obtains the names and addresses of employees in the activity in which he has an interest. Perhaps a small bribe is used to obtain the data. He then designs an innocuous-appearing questionnaire that resembles others that the employees receive routinely. The difference is that the spy has inserted cleverly worded questions that, if answered, will give him the information he wants. He may send the questionnaire to all employees in the facility or only to certain selected individuals whom he knows have the information he is after.

Usually the spy uses a cover so he cannot be identified if an attempt is made to track down the sender of the questionnaire. He may arrange with another organization to conduct the survey for him, or he may use a fictitious name and a post-office box for the return address. If the questionnaire results are not all that were expected, the answers given may at least help the spy determine what he should do next to obtain the information that is missing. For example, the survey results might reveal an individual in the facility who could be developed into a patsy.

Mergers, Acquisitions, Joint Ventures, and Licensing

A proposal for a merger, acquisition, joint venture, or licensing arrangement has been used to obtain trade secret data. The company that makes the offer in such a situation usually makes the terms so attractive that the victim company is unable to resist negotiating. As a part of the negotiation the proposing company asks for information not only about the other company's patent holdings and applications but also about innovative ideas. Further, it wants complete records of sales, price structure, future plans, suppliers, customer lists, and other sensitive data. And after it has collected all the available information, it terminates negotiations.

A number of companies have been victims of such negotiations and so have become careful about the type of information they are willing to disclose in any situation.[10] But being too careful might also be detrimental, because an opportunity to capitalize on the offer of an earnest negotiator might be lost if sufficient information is not provided.

Some companies have solved the problem of the protection of proprietary information by dealing through an intermediary. Detailed information is not given to the company that asks for it. Instead, an independent party such as a consultant or an attorney is allowed complete access to records under detailed restrictive conditions. Sufficient information can then be transmitted to the representatives of the proposing company without exposure of proprietary information and trade secret data of the company being contacted.

Applicant Interview and Résumé

Job applicant interviews and résumés have been used effectively to collect information from competitors. A company that is interested in a particular kind of information advertises a nonexistent job opening in that field. The objective is to attract a number of individuals who work for competitors as applicants for the position.

The applicant who responds to the advertisement is interviewed in great detail by an expert who is interested only in obtaining as much information as possible about the proprietary data of the applicant's company. If the interview indicates that the applicant has information of value, he is asked to develop his qualifications more fully in a résumé or on a company form supplied by the interested organization.

The company conducting the feigned search can, of course, make the nonexistent position most attractive and promise every possible benefit to whet the applicant's appetite for it. In such a competitive situation an applicant is not only anxious to demonstrate his capability by explaining his technical knowledge but also eager to be completely cooperative. Because very sensitive data can be lost in this way, the hazard should not be overlooked when the industrial espionage prevention program is developed.

Visitors and Customers

A visitor must, of course, always be treated courteously. At the same time, his activities must be controlled because he may be a representative of a competitor who is searching for information or he may be an industrial spy. A knowledgeable visitor who knows what to look for can reap a rich harvest of information if his movements in an enterprise are not limited.

Plant tours are recognized as important from a public and community relations standpoint, but a number of companies have recognized that a complete plant tour can reveal a great deal of information about processes and activities. Because of that hazard, some companies have eliminated tours entirely and others have restricted them to areas from which little or no information can be obtained.

Several years ago a visit to one of the leading electronics manufacturing facilities by a group of foreign business leaders resulted in the loss of a secret manufacturing process that enabled competitive firms in that foreign country to begin to compete with the U.S. firm. A complete tour of the manufacturing facility guided by a flattered top company executive is blamed for the loss.

A vice-president of manufacturing in one of the largest food-processing companies has said that he can walk through a competitor's plant and, by observing the processes and the number of people employed, estimate almost exactly the production figures for the plant. On one occasion when his company was negotiating a merger with another company, he went to the plant as a visitor and quickly determined after a short tour that the management of the organization had falsified production and sales data. As a result, the merger negotiations were stopped.

Reverse Engineering

"Reverse engineering" is a common industrial term for the complete analysis of a product by a competitor's technical personnel to the end of copying, changing, or improving the product. If the product is patented, it may be necessary to "design around" the patent. Enough changes are made in the product so there is no patent infringement or so a new patent can be obtained by the company that designs around the original one.

Reverse engineering is widely practiced. It is completely legal and is regarded as ethically acceptable. It has been reported that the first cars off the Detroit assembly lines are obtained by competing manufacturers so they can be completely torn apart and carefully analyzed. If an organization is able to use reverse engineering successfully, it can, of course, avoid licensing and the payment of royalties.

It is impossible to prevent a competitor from using reverse engineering on a product that is sold on the open market. Then, however, the practice is less harmful than it would be if applied to a prototype or model still in the development stage. If a competitor can take advantage of reverse engineering before a product is marketed, he may be able to gain a great competitive advantage. The lead time normally expected by the originator might be lost, and the competitor might be able to get his copy of the product onto the market first. Therefore, the industrial espionage prevention plan should stress the protection of products not yet in production.

Subcontractors

It may be necessary to send material containing proprietary information out to a subcontractor who is engaged for the fabrication of models, components, or parts, for reproduction and printing, or for the computer processing of data. When such material is in the subcontractor's facility, it is, of course, no longer under the protection of the organization that originated it. Regardless of the effectiveness of an industrial espionage prevention program in the originator's facility, if the subcontractor's facility does not have an effective program that it agrees to use in the protection of the entrusted data, the material will be vulnerable to all of the hazards discussed so far.

The subcontracting company might be ethical and be above knowingly compromising customer trade secrets. On the other hand, it might not be aware it is dealing with trade secrets, and it might not have an effective security program in operation. As a result, valuable trade secrets might be lost while in the hands of the subcontracting company simply because neither the value of the information nor the reality of the industrial espionage threat is appreciated. The industrial spy who wants a particular type of data can be expected to make it his business to know that the material is being sent to a subcontractor. He will soon find out that the material is not being protected, and he will arrange to obtain it there. In that way he will circumvent what might be extensive and costly protective measures at the prime facility.

REFERENCES

1. *Business Week,* October 6, 1962, p. 66.
2. "Industrial Espionage: Operation Air Bubble," *Newsweek,* April 19, 1965, p. 79. Used by permission.
3. "How Russia Spies: The New Game," *Newsweek,* October 11, 1971, p. 31. Used by permission.
4. "How I Steal Company Secrets," *Business Management,* October 1965, p. 3.
5. Ibid., p. 5.
6. Ibid., p. 6.
7. Richard Austin Smith, "Business Espionage," *Fortune,* May 1955, p. 118. Used by permission.
8. John Perham, "The Great Game of Corporate Espionage." Reprinted by special permission from *Dun's,* October 1970. Copyright, 1970, Dun & Bradstreet Publications Corp.
9. *The Wall Street Journal,* August 8, 1970. Reprinted with permission. *E. I. du Pont de Nemours & Company* v. *Christopher,* 166 U.S.P.Q. 421, 423-425 (1970).
10. *The Wall Street Journal,* June 28, 1968.

6

---◆---

BUGGING AND WIRETAPPING

"Would you believe your office may be 'bugged'? You'd better believe it." That is a quote from the front cover of a sales brochure distributed nationally in 1971 by an organization offering to "perform technical security surveys in your plant to guard against electronic eavesdropping and wiretapping on a one-time or continuous basis." The front cover text continues:

A recent survey revealed that despite the 1968 Statute out-lawing electronic eavesdropping, these devices are now readily available and at lower prices than ever before. For instance, a non-technically trained person can buy a $10.00 device in a local store, plant it in your office in less than 2 minutes, and listen to all of your private conversations from the safety and comfort of his car parked down the street or from a nearby building. His risk of arrest is almost nil since he is exposed only while planting the device, and if the device should be discovered, it is very difficult or impossible to trace to the offender. There are also such terrifying "Big Brother" devices which can be purchased, no questions asked, as the so-called "infinity transmitter." This insidious device, when attached to your phone, allows the culprit, by calling your telephone number from anywhere in the country, to listen to all conversations in your office or board room without your knowledge. (Cost: about $200, installation time about 20 minutes, no technical training required.)

A frightening prediction of what can happen? Of course it is, and although the techniques described may be oversimplified, the threat

68

is real. Because of the extensive publicity given to the subject of electronic eavesdropping in the 1960s, it may be difficult to separate fact from fiction. The olive in the martini glass captured the imagination of the news media as well as the public a few years ago, and it was probably given more attention nationally by the news media than any other espionage technique in recent years. Television and movie stories that portray the use of electronic listening devices, sometimes in completely unrealistic, impractical, or even ridiculous ways, have also brought the subject to the attention of the public.

There are experienced security executives in business and industry who maintain that the use of electronic listening devices is stressed far out of proportion to the problem it presents. They point out that there are so many other means of obtaining information, as was indicated in the preceding two chapters, that electronic bugging is not an important threat. But despite the variety of opinions and information, sometimes misleading and inaccurate, about electronic listening techniques and devices, a realistic defense against electronic snooping should be included as one element in the complete industrial espionage prevention plan.

Some recorded instances demonstrate the practical value of information that can be secured through electronic eavesdropping. Hazel Bishop, the cosmetics firm, reportedly lost $30 million in the mid-1950s because of wiretapping. A competitor arranged to bug the office of the president as well as the executive conference rooms in the company's New York headquarters. The competitor representatives were able to listen in on conferences at which plans for new products were being discussed. The information obtained allowed the competitor to reach the market as much as six weeks ahead of Hazel Bishop with identical products.

A $50,000 reward offered by the head of a company in the 1960s for information leading to the identity of the individual who had bugged his telephone was a clear indication of the value of information that had been obtained. Matthew Devine, Chairman of the Amphenol Corporation, an electronics manufacturing company in Oakbrook, Illinois, became suspicious when confidential information about the company was being obtained by outsiders while the company was involved in merger and acquisition negotiations. In the course of an investigation, a listening device was found on a telephone in Mr. Devine's home.

Perhaps the Amphenol incident established $50,000 as the amount that should be offered as a reward in such cases, because another $50,000 reward was advertised to readers of the *Los Angeles Times* on October 10, 1971. One of the authors of this book sent a letter to the box number given in the advertisement. He said that this book was then in preparation and that information regarding the incident that motivated the ad would probably be of value as an example to be used in this chapter. As might be expected, no reply was received, and so the incident that caused the reward to be offered is still a mystery.

Prevention, which has been stressed throughout this book, is the key to the problem of the threat of electronic eavesdropping. It is now a federal offense for an individual to use, possess, manufacture, or sell eavesdropping devices. Although the violator of the law can be arrested and charged with a crime, that will be of little satisfaction to the victim if information already obtained has caused serious damage. The legal aspect of electronic listening will be developed in more detail later in this chapter. Now, however, the reader is cautioned not to depend entirely on the federal and state statutes for protection against the threat.

Electronic Listening Techniques

Bugging or wiretapping is actually relatively easy for anyone who has had some technical training or experience. However, the discussion in this chapter is not intended to be technical. Instead, it is a general nontechnical outline of some of the more common techniques of electronic listening. For readers who desire additional detail, a number of books that deal with the technical details of electronic eavesdropping are available.[1]

An electronic listening unit is generally composed of three elements in addition to a power source: a microphone to gather information, a means of transmitting the signals to a remote location, and a listening post at the remote location to receive the information gathered. The transmission of information is usually accomplished by wire or by radio. Wires may be installed to connect the microphone with the listening post, or existing wires may be used—telephone lines, intercom lines, alternating-current power lines, burglar alarm lines, and so on. At the listening post, earphones, speakers, or tape recorders may be used.

MICROPHONES

A microphone to pick up conversations or sounds may be specially installed in the area to be monitored. Alternatively, equipment already installed, such as a telephone, radio, television, or intercom speaker, may be converted into a microphone. Of the many microphones that can pick up signals, some are so small they can be concealed in lamps, in picture frames, under carpeting, or in any of the variety of items commonly found in a room. If a microphone is installed by an experienced technician, it may be very difficult to locate.

Because many types of microphones are available, no attempt will be made to discuss all of them. Instead, a few of the more common types will be discussed so that the reader who has limited knowledge in the area will be better able to appreciate how easy it is for experienced personnel to install electronic eavesdropping devices.

A very common microphone utilizes carbon granules suspended between a diaphragm and a retaining plate. Sound waves that strike the diaphragm cause it to oscillate, and the oscillation alternately compresses and releases the carbon granules. The varying density of the carbon modulates an electric current passing through it, so that the characteristic changes, when later demodulated, reproduce as the original sound waves. Very early radio buffs may remember that announcers and other studio personnel used to tap or shake the carbon microphones to break up impactions in the granules. Today the carbon mike, although still in use, is far exceeded in performance and reliability by other devices.

Other microphones include the crystal, ceramic, and dynamic types. The first utilizes mineral crystals instead of carbon granules. The crystals not only modulate the sound waves but also develop the electric current necessary to carry the signal. Crystal mikes, however, are highly unstable in their response to changes in temperature and humidity, and they can be used reliably only in a controlled environment. Ceramic microphones have slices of ceramic material rather than crystals but otherwise operate in a similar way. They are less fragile, but they also produce less satisfactory results.

The dynamic microphone operates as a loudspeaker in reverse. It involves a diaphragm surrounded by a wire coil suspended within an electromagnetic field. Sound waves that strike the diaphragm cause it to oscillate within the magnetic field, and that produces a variable electric current along the leads A variant of the dynamic mike is the velocity mike, which uses a metal ribbon or strip instead of the diaphragm and wire. Microcircuitry and miniature electronics can now produce devices with microphones operating on these principles in housings as small as a wristwatch, a fountain pen, or a hearing aid.

Still another type of microphone is one that senses vibrations or low-frequency sounds that result when the spoken word strikes a wall or like surface. It has been regularly used in such places as adjoining motel rooms when direct access to the target is not possible. The microphone consists of a pointed sensor, or spike, and related wire leads. The spike is inserted into a predriven hole to a point just short of the surface of the opposite wall. There, although it is not visible to the target, it picks up all the sound waves that strike the wall in the target space. Of course, the drilled hole will leave evidence in the room in which the mike is used. Enterprising espionage agents have carried plaster of paris and other instant repair kits to disguise even that minimal damage.

It is interesting that early legal decisions in the eavesdropping cases made distinctions between spike mikes that involved a trespass by entering the area or space of the target and others that merely picked up the freely emanating sound waves. The trespass concept was familiar in common law and also in most criminal statutes, and so it provided a familiar basis for distinction. Simple listening posed a more difficult problem—one more illustration of the impact of technology on ethics.

Another exotic device features flexible acoustical tubing that can bend around corners or over transoms. The common stethoscope can be and has been adapted to listening by connecting it to an amplifier. The parabolic mike, which features a sound guide in the form of a parabola, will concentrate distant sound into a microphone and has been used to eavesdrop for considerable distances.

TELEPHONE BUGGING

The bugging of telephones has been a popular means of electronic listening for two reasons. First, telephones are readily available in many locations; second, the technique is simple for an individual with a minimum of basic technical training.

The easiest way to bug a telephone line is to use an induction pickup or coil. In its basic form, the coil will pick up conversations if it is placed alongside the telephone. The telephone can also be wired so that it operates as a microphone when the handset is in the cradle or will operate only when the handset is lifted.

In addition to induction pickup bugging, an experienced technician can tap directly onto a telephone line. Unless such a tap is properly made, however, a meter placed on the line will indicate an unusual load and the tap will be revealed. Nevertheless, an individual who has been properly trained can make an installation that cannot be easily discovered.

A technique that was given a great deal of publicity in the 1960s involves the use of an infinity microphone. With that device an individual can wire a telephone so that it will become an open microphone on signal. The eavesdropper can then direct-dial that telephone's number and, immediately after dialing the last digit, actuate the infinity microphone in the telephone instrument with a whistle or harmonic note. The phone will not ring when the whistle or harmonic tone is interjected onto the line. The infinity microphone excited a great deal of interest because it can be activated from thousands of miles away. An individual in Los Angeles, for example, can bug a room in New York. The infinity microphone will remain active as long as the eavesdropper keeps the line open by having his receiver out of the cradle.

MINIATURE TRANSMITTERS

The bugging devices discussed up to this point have involved the use of wires to transmit signals, but radio-frequency transmission is more popular than wiretapping because it requires less technical knowledge. No expertise with telephone systems or wiring systems is required; a radio link cannot be traced; and the radio bug is considered expendable because of its relative low cost. However, radio bugging is usually not as dependable as wiretapping, and the recordings are seldom as good as those obtained by wiretaps.

It was mentioned earlier in this chapter that the television industry has helped to publicize the use of electronic listening devices. Television has also contributed to electronic listening in another way. The miniature transmitters that are now used in radio bugging were developed in the 1950s to meet the needs of television studios. The transmitters were wanted so that microphone cables, which were a safety hazard, could be eliminated. As a result, a great many different types of miniature FM radio transmitters that could be concealed on the person were developed, and those devices have become favorites for radio bugging.

A radio bug consists basically of a microphone, an oscillator modulated by the microphone, a power supply for the oscillator, and an antenna to transmit the modulations. Of course, a receiver must be provided at the listening post to detect and demodulate the broadcast signal. A miniature radio transmitter can be easily concealed in a room or area to be bugged. In fact, several devices might be installed to insure good-quality reception.

A radio transmitter is often combined with a telephone, which can be bugged in several ways. One type of tap is designated the drop-in: the eavesdropper installs a miniature FM transmitter in the plastic mouthpiece of the telephone handset. The device does not require any batteries because it draws its power from the telephone company supply. The transmitter will operate when the telephone is out of the cradle, and both sides of a conversation can be monitored. The range of such a transmitter is short because of the limited power available.

Another type of telephone tap, referred to as a parasite transmitter, also uses the telephone power supply. It is installed in the telephone instrument in the handset, along the line, or at the terminal. The device will transmit only when the handset is lifted, and its range is up to one-quarter mile.

The parallel transmitter differs from the drop-in and parasite types. It is normally not connected to the telephone instrument itself but is instead connected to the wires leading from the phone at a terminal box or on a pole. The tap is self-powered and can transmit up to one mile under favorable conditions.

A radio transmitter requires power, and so its size will depend to some extent on the power unit provided. Transmitters are versatile and can be concealed on the person as well as in an area or on a telephone to intercept and transmit conversations to another location.

Detection of Electronic Listening Devices

It has been estimated that $100,000 worth of equipment and a technician with at least ten years' experience would be needed to locate every possible eavesdropping device. Some experts might argue that the equipment cost is too high, but few would dispute the need for technical skill. An experienced operator will include in his search both a physical inspection and an electronic inspection of the facilities. The physical inspection will include a careful analysis of the telephones in the area to insure that they have not been bugged, as well as an inspection of other areas in which bugging devices might be installed.

The electronic inspection will normally include the use of a metal detector to locate microphones or wiring. However, the detector cannot be relied on completely, because the concealed device might be constructed of a material that it would not readily pick up. Other devices can be used to check telephone lines to determine if listening equipment is installed on the lines. Again, the equipment is not completely reliable because the telephone installation can be such that the presence of listening devices will not be detected.

The experienced operator will also be expected to use radio-detection equipment to determine if there are any radio transmitter bugs. Since the transmitters may operate on any of a variety of frequencies, a considerable amount of equipment is necessary to search the complete spectrum of frequencies. Unfortunately, an operator may not be able to detect a radio transmitter that is not operating. As a result, a transmitter that has been designed to be operated by remote control or has been programmed to be turned on at a particular time and is not operating when the inspection is made will probably not be found.

THE PHYSICAL SEARCH

No matter how advanced the electronic search gear in use, every detection inspection should include a comprehensive physical search. The physical search is necessary to locate devices that are not operating anywhere in the r-f (radio-frequency) spectrum at the time the electronic sweep is made. They could include not only radio devices that are inactive at the time but also wire devices that might in fact be operative. Because of the low power requirements, the miniature nature of the microphones and amplifiers or transmitters, and the extremely fine wire that is available when wire is needed, it is extremely difficult, even when a comprehensive normal physical search is made, to be sure devices have not been missed.

It is not unusual for searchers to remove moulding or paneling, take up rugs and carpets, remove gratings and grilles, partially disassemble furniture and fixtures, and trace all wiring, electrical and communication, that enters the space being searched. Obviously that requires time; and in large conference rooms or very ornate offices, the search may take an entire day or longer.

Because the space searched is apt to be in regular business use, the occupant will want it restored to the presearch condition. That is noted because some persons are under the illusion that a dependable countereavesdropping inspection can be made quickly by passing several sensing instruments over walls and around telephones. Although it might be possible to find an "expert" willing to perform so perfunctory a check, the investment in money and inconvenience would not be prudent. The physical search is an integral part of the inspection. In fact, even if the instrument-detection effort is successful, the device located will still require physical removal.

THE ELECTRONIC SWEEP

The electronic sweep relies quite heavily upon feedback when devices operating in the r-f range are sought. Feedback is the squealing noise experienced when a transmitted sound is picked up by the microphone that first transmitted it and is then amplified, transmitted, picked up again, and so on in an endless chain of constantly increasing amplification of the same sound. Most readers will have experienced feedback when the gain or input volume of the public-address system at an assembly function has been too high. The speaker begins and a low whine rapidly becomes a piercing shriek until he stops or the gain is reduced. It is exactly that sound behavior that makes possible the finding of active transmitting devices.

In a search for a device, sound is introduced into the space being searched. An AM-FM radio is often used. As each section of the target space is inspected, the sound is transmitted over a different frequency, as by changing stations. With a very advanced detector, an extremely broad r-f band can be sampled in that way. A typical FM radio alone will transmit over all the frequencies between 88 and 108 MHz. If the hidden device is r-f, is operating at the time of search, and is also in the 88- to 108-MHz range, the feedback squeal will be heard at the moment the FM radio signal is transmitted on the same frequency as the device. As the radio is tuned away from that frequency, in either direction, the squeal will stop. By zeroing in on the exact frequency and keeping the feedback going, the searcher can locate the device by the volume changes as the tuner nears the location.

Of course, if the frequency of the hidden device is not within the 88- to 108-MHz width (and it probably won't be), the device cannot be located with FM radio feedback. That explains why so much gear is needed for a thorough job—many frequencies from low to ultrahigh must be checked.

The requirement that the hidden device be operating is most important. Often the area to be searched is in constant use during the business day. Concerned management may want a complete antieavesdropping inspection but want it over the weekend when the space is not in use. That the space is not then in use may be exactly the reason why the espionage agent would select a device that can be switched on and off remotely. A weekend search will then fail because the device is not operating. However, as their unit costs go down, more radio-transmitting devices are regarded as expendable and are left to operate continuously until they break down. That would make detection just as easy on a weekend as at any other time.

The search for devices that do not transmit r-f signals will have to depend upon physical discovery as discussed earlier in the chapter. However, if absolute security is desired, it should not be assumed that any search has been completely successful. Countermeasures should be employed to defeat the listening device that may have been missed.

COMMONSENSE PRECAUTIONS

The best countermeasure is not to hold the conversation in the place that may have been bugged. The last-minute movement of a business meeting from an announced location to a different one may be enough to upset the most ambitious espionage plan. If the location cannot be changed, the introduction of masking noise, although something of an inconvenience to the legitimate business, may defeat surreptitious listening. A radio, an electric fan (especially one with paper or other foreign object inserted to be struck by the blade), or even multiple conversations may neutralize a clandestine listener. If the masking noise is introduced between the target conversation and the bug, it will be most effective. Although such simple precautions are admittedly home remedies, at times they may be the only ones available, and they should not be overlooked.

If the telephone is suspect—and a telephone in a sensitive location should always be suspect—one solution is to use a jack instrument rather than a prewired one. If a stock of one or two spares is maintained under tight security control, the instruments can be interchanged on an irregular basis. The instrument last removed is subjected to close bench examination. The mere regularity of change would probably convince a spy not to use the telephone. Even multiple-extension or button instruments can be handled like jack types by using fast-connect cable couplers. The added expense for the precautionary measure may be offset by avoidance of losses in the hundreds of thousands of dollars.

Laws Against Eavesdropping

THE FEDERAL LAW

In its report *The Law and Private Police*, The Rand Corporation notes that, between 1958 and 1968, sales of electronic detection and surveillance equipment grew from $27 million to $83 million.[2] In that same period, various states had tried to deal with the eavesdropping problem on a local statutory basis. However, no effective federal criminal sanction existed, and the provisions of the Federal Communications Act of 1934, although they could be and were used to suppress evidence in federal criminal trials, did not spark any large number of criminal prosecutions against violators. The Rand study notes that, between 1934 and 1969, there were fewer than twenty federal prosecutions for wiretapping.

The Omnibus Crime Control Act of 1968, however, contained broad and effective federal prohibitions. Codified as 18 U.S.C. 2510 and following sections, the law basically provides as follows.

1. It is a felony punishable by five years' imprisonment or $10,000 fine, or both, (a) to willfully intercept, try to intercept, or procure another to intercept or try to intercept any wire or oral communica-

tion or (b) to use or try to use or procure another to use or try to use any electronic, mechanical, or other device to intercept any oral communication when such device transmits a signal by radio or interferes with radio transmission or when it is affixed to or transmits a signal through connections used in wire communications, or when the user knows that the device or any part has been transported in the mails or in interstate commerce, or when the use takes place within or develops information relative to the operations of a business whose operations affect interstate commerce.

2. It is also a five-year, $10,000 felony to disclose willfully or to attempt to disclose the contents of any wire or oral communication that the discloser knew or had reason to know was obtained through interception in violation of the section.

3. The use or attempted use of the contents of any wire or oral communication known to have been procured through illegal interception is made a felony with punishment as in (1) and (2).

4. Manufacture, assembly, possession, sale, and transmission through mails or in interstate commerce of any electronic, mechanical, or other device whose design renders it primarily useful for the purpose of surreptitious interception of oral or wire communications, with knowledge or reason to know of such primary purpose and of the past or future interstate or mail transmission, are prohibited, as is advertising any such device with knowledge of its primary purpose or of any device in such a way as to promote it for such purpose. The penalty in these cases is also five years and $10,000.

5. The Act exempts from coverage switchboard operators or other employees or agents of communication common carriers who are acting in the normal course of employment to intercept, use, or disclose communication if the employment activities are necessary to rendition of the carrier's services or protection of the carrier's rights or property. It also exempts FCC personnel acting pursuant to the Communications Act and private persons who are intercepting communications to which they are a party or to whom one of the parties has given consent prior to interception. Thus anyone can intercept and even record his own wire or oral communications. But that does not mean that someone who is the listed subscriber to a telephone, for example, can legally intercept a call on that phone made to or by other parties. The Act is concerned not with the communication instrument but with the privacy of the par-

ticular communication. Another point in the same connection is that although a party to a communication may intercept it without violating the interception prohibitions of the Act, he may not use an instrument primarily intended for the surreptitious interception of wire or oral communications. That could be as practical a distinction as whether an intercepted telephone conversation is recorded on an ordinary tape recorder or a recorder specially modified or designed for one of the uses discussed earlier in the chapter.

In addition to the criminal penalties, the federal Act also creates a civil right of action by an aggrieved person against the one intercepting, using, or disclosing the communication. Damages are actual damages (not less than the higher of $1,000 or $100 per day for each day of illegality), plus exemplary damages when malice can be shown, plus costs and attorney's fees. The Act also bars receipt in either a federal or state court of any evidence received from an improperly intercepted communication as well as the contents of the communication itself. That section applies irrespective of who committed the illegal act of interception—government agent or private citizen.

The federal statute does three things that should help reduce industrial espionage by electronic devices and eavesdropping. First, it makes such behavior a federal felony. Second, it makes the knowing manufacture, possession, sale, or advertising of espionage devices a federal felony. That should shorten the supply of specialty items because it exposes everyone in the supply chain to prosecution. Third, it gives aggrieved persons a special right to civil damages. The civil right is in addition to the other civil remedies available to a victim of industrial espionage. Those remedies are discussed in detail in Chapter 3.

STATE LAW

A number of states had passed statutes of their own before the passage of the Omnibus Crime Control Act, and others have passed or amended statutes since. The federal law does not exclude action by the states, and state laws may forbid things that the federal law permits. In California, for example, no one may record a telephone conversation without the consent of all the parties to it. The federal law, and also the New York statute, permits the interception of a wire communication with the consent of a single party.

THE PRESENT SITUATION

Although federal and state laws make it a felony to eavesdrop, the advance in technology makes eavesdropping increasingly easy. When the target is big enough, the sanctions of the criminal law will not deter an aggressor. It is obvious that, if they did, there would be less, and not more, crime.

The prudent entrepeneur will continue to take all practical steps to protect his sensitive information against electronic as well as physical theft. He will also remember that, when his counter-measures do expose an aggressor, powerful new penal laws and civil remedies are available. He should certainly make maximum use of them.

REFERENCES

1. Robert M. Brown, *The Electronic Invasion* (New York: John F. Rider, Publisher, Inc., 1967). John M. Carroll, *The Third Listener* (New York: E. P. Dutton & Company, Inc., 1969). Samuel Dash, Richard F. Schwartz, and Robert E. Knowlton, *The Eavesdroppers* (New Brunswick, N.J.: Rutgers University Press, 1959).
2. *The Law and Private Police*, Vol. 4, Report R-872/DOJ (Santa Monica, Calif.: The Rand Corporation, 1971).

7

---◆·◆---

REDUCING THE RISK–
A SYSTEMS APPROACH

The approach to reducing the risk of information loss must be broader and deeper than the approach to any other security problem because many more people may have unsupervised opportunity to cause an information loss. In breadth the coverage must include all elements of the organization—information is exposed in all locations, and not merely in the relatively few departments and activities that generate sensitive data. In depth the coverage must be such that every employee, right down to the newest line recruit, is properly prepared to recognize and discharge his unique obligations.

The proper place to begin an effective information control program is with a statement of general policy published at the highest level of the enterprise and circulated as widely as possible. If policy distribution is normally restricted to a limited number of executive personnel, it will be necessary to prepare special materials to carry the policy message to the entire workforce. The materials can take whatever form is most useful in the given enterprise: procedure statements, addenda, local implementation instructions, or even specially prepared pamphlets or booklets. Sometimes the booklet technique will be indispensable, particularly if the quantity of sensitive data is large, the exposures are varied, and the employee population is of considerable size.

THE POLICY STATEMENT

The basic objective is to make a statement of policy. It is not uncommon to find a purported statement of policy on information control that begins with procedure and never gets to policy at all. The following is a typical simple statement of policy:

Company information will be protected by measures commensurate with its nature. Safeguards will be established to prevent disclosure of information which would be harmful to the company. No employee will disclose any company information to persons outside the company unless that information has already been made public or approved for publication. It is the personal responsibility of an employee desiring to make public disclosure to be certain the information has been approved for publication.

The preceding brief statement of policy does three things: (1) It says that there will be a formal program of information control. (2) It establishes the standard for information control prevention: *harm to the company.* (3) It assigns each employee, personally, the responsibility for being sure that any information he releases to any person outside the company has been approved for disclosure. Much elaboration may follow in the statement of policy itself and in related procedural instructions, but the basic tone has been set in one short paragraph. The most troublesome information disclosures, those to persons not in the company, have been specifically identified as requiring prior approval, and each employee has been made individually responsible for refraining from making any such disclosure himself without the requisite approval.

Of course, a mere statement of policy will not itself prevent employees from making unauthorized disclosures either inside or outside the company. However, the employees of goodwill—the majority in most companies—will have a rule that they will obey, given reasonable opportunity to learn it and to apply it in practical situations. Now the duty rests with the enterprise to make effective procedural control possible by developing the right techniques and educating the workers in the proper application of those techniques.

DEFINING SENSITIVE INFORMATION

The next step is to identify the kinds of information that are sensitive and indicate how they may be recognized. Earlier discussions have emphasized that trade secrets are only one kind of sensitive information. Financial, organizational, and personnel information will also be sensitive from time to time; indeed, in some companies the greater volume of sensitive data will fall into those classes and not into the trade secret class. In any case, an overall label or set of labels is required to identify any data that meet the standard set by the policy, that is, information the disclosure of which would be harmful to the company. The labels are found by applying a principle that has long been familiar to businesses dealing with the military and security agencies of the government: the principle of classification.

In application of the principle of classification, information is divided into two or more categories one of which includes all information that is not sensitive and requires no special precautions and the other or others of which contain all sensitive information. Almost everyone is acquainted with the designations TOP SECRET, SECRET, and CONFIDENTIAL, which are used by the U.S. government in its information-safeguarding program.[1] Information in those three categories, whether technical, financial, operational, or whatever, is sensitive, and its disclosure would be harmful *in differing degrees* to the defense effort. The degrees of sensitivity are explained in the definitions of TOP SECRET, SECRET, and CONFIDENTIAL. The definitions themselves are not important to this discussion, because they will not be precisely applicable to purely industrial or commercial situations. What is important is that the definitions do distinguish classes or categories of data all of which are sensitive but some more so than others.

The vast amounts of data with which the government deals in the defense program obviously require elaborate precautions, including the development of multiple categories of classification. However, the more complex and elaborate the control program becomes the less its chances for optimum effectiveness. An industrial information security classification system should borrow the principle from the government but not necessarily the practices. Interestingly, many industrial security executives have argued for years that there are too many classifications even in the government program. As long ago as 1957 a presidential commission, after exhaustive study of all the information security activities in all the executive departments, recommended the abolition of the classification CONFIDENTIAL as being unnecessary and costly.[2]

Some businesses may feel a need for two or more classifications of sensitive information, but in general it is best to keep the number of classifications as low as possible and avoid subtle distinctions. For example, some kinds of information processed by the company will, if released, embarrass or conceivably even harm individual persons but not harm the enterprise itself. Personal data about employees are the best example of that kind of information. And although precautions should be taken to avoid disclosures of personal information in ways that are discomfiting or prejudicial to the affected employee, it is important to distinguish between such disclosures and those that would be harmful to the company. There have been instances in which highly formalized information security programs were successful in protecting personal data about employees and unsuccessful in protecting vital data about the enterprise. It is easy to promote concepts of security programs that are so inexact that the workforce will have an immediate protective response to personal data and fail to protect highly important technical data.

DEFINITION AND CRITERIA

A general standard—harm to the company—has already been suggested as the platform on which to build the protection program. It is equally important to establish specific criteria or bench marks by means of which information can be evaluated as meeting or failing to meet that standard. Criteria will be most useful if they suggest practical tests with which to measure given data. Information that exhibits the qualities of or is identifiable with one or more criteria will be designated sensitive and require protection.

To communicate the judgment that specific information is sensitive and must be protected, there is need of a set of designations that will perform the function for the industrial proprietor that TOP SECRET, SECRET, and CONFIDENTIAL perform for the government. However, it is extremely important that the government's terms be avoided in the purely proprietary program. Government classified information must be unmistakably separated from proprietary classified information to assure that each is protected as it should be. Firms that perform classified work for the military and other government departments and have agreed to be bound in that work by one of a variety of security agreements find that an intensely practical requirement that may be faced on a daily basis.

Failure to protect classified defense information can have serious civil and even criminal consequences, but effective proprietary programs can usually operate on less detailed procedures than government programs can. For example, incorrect labeling of certain information could result in its being protected adequately from a company point of view but inadequately from a governmental point of view. First-class U.S. mail might be and often is a satisfactory transmission medium for proprietary data, but is not permitted for classified defense data. If both company information and defense information were classified SECRET, lapses and violations could easily occur.

It has been suggested that adding the word "company" to the label, as by marking data COMPANY SECRET or CONFIDENTIAL COMPANY DATA, would solve the problem of confusing proprietary and defense data. In part that is true, but only with regard to documentary information. With regard to oral disclosures, it is less likely that persons will remember the classification as being proprietary rather than defense, or the reverse, if any of the same words are used in the labels. If lack of inventiveness is the only limitation, there is no good reason why companies cannot choose classification labels that are totally distinct from those used by government. The following is a partial list of titles already in use by companies in the United States:

Limited	Private
Proprietary	Company only
Internal	Restricted
Registered	Controlled
Sensitive	Secure data

By taking the term "proprietary" from the list for illustrative purposes, we can now develop the definition of sensitive information to be used in the policy statement. In a section of that statement containing *all* special definitions or unique terminology applied to information control, the following could be included:

PROPRIETARY. A classification assigned to any Company information whose unauthorized disclosure could result in harm to the Company or be detrimental to Company interests, and which must be safeguarded in accordance with this policy and supplemental instructions.

This specimen definition incorporates the standard established earlier and says that any data so defined will be subject to the controls on handling contained elsewhere in the policy or in related instructions. It is not yet indicated how such information will be identified, but it is now clear that any information that has been so identified requires special safeguards that are to be found or referenced in the policy.

To assist those who will have the responsibility of making original determinations of just what information is proprietary, a set of criteria should be developed to reflect the particular nature of the enterprise. Such criteria can suitably consist of lists of the kinds of information that are likely to be or contain proprietary data. The following are examples of such criteria taken from the working policies of various organizations:

RESEARCH AND DEVELOPMENT INFORMATION. Any information relating to company-sponsored research or development projects.

FINANCIAL INFORMATION. Financial data may include such items as pricing of company products, operating statements, return on investment, profit margins, and bid or proposal quotations.

TECHNICAL INFORMATION. Technical information includes product specifications, process specifications, test results, formulations, and performance characteristics and limits.

MARKETING INFORMATION. Marketing information includes schedules of delivery or distribution, customer lists, assessments of market and share of market, pending or planned negotiations, customer analysis, and reports on products.

ORGANIZATIONAL INFORMATION. This heading covers information regarding opening, expanding, or modifying facilities, mergers and acquisitions, and transfers of responsibility or personnel.

These criteria are not exhaustive, but they are generally representative and would apply to most commercial organizations. Each criterion should be developed for the particular enterprise that will use it. The most effective approach to that development is to establish a consensus among senior managers of all the company functions. In a typical organization the following group might represent senior functional management: president, vice-president research, vice-president finance or treasurer, vice-president marketing, vice-president manufacturing, general counsel, and vice-president engineering. Each of the executives will have a profound insight into the kind of data generated by his part of the organization and will be able to specify the data likely to constitute proprietary information. And each will also have some ideas about the sensitivity of the other functions as well.

Consensus among senior functionaries should assure that the really important information assets of the enterprise have been identified, but arriving at a consensus may be neither easy nor quick. Often an executive will have an exaggerated sense of the importance of certain information. However, by making the initial determination of criteria at the highest level, a good deal of the limited perspective that sometimes results from sharp departmental or functional boundaries at lower levels can be avoided. To put it another way, if the senior functional management, as a group, is not best qualified to agree on what information assets the company really has, it would be difficult to say what other group is qualified.

Note that the consensus is *not* on a particular piece of information, although examples of data will certainly be used in arriving at consensus. It is, rather, agreement on the kinds or families or categories within which sensitive information will occur. Once that agreement has been reached, the responsibility for deciding about individual pieces of information can be assigned much lower in the organization. That is discussed a bit later.

A very important advantage to be gained by establishing consensus among the senior group as an early step in the control program is to assure later understanding and willing participation of the group's members in the execution of the detailed control procedures. In very large organizations the effect on subordinate groups and departments will be far more favorable if it is known that the program was developed by the senior managers themselves.

THE FUNCTION OF CLASSIFICATION

The single most important decision in the entire control program is the decision to classify, and that is one that will be made repeatedly, perhaps dozens of times daily. It is critical for three reasons. First, if truly sensitive information is not recognized early enough, there will almost certainly be unauthorized or at least undesirable disclosure. It was noted earlier that such disclosure can forever foreclose the original proprietor of the information from taking protective measures if innocent persons change their positions in reliance on the disclosed data.

Second, essentially nonsensitive data can be improperly classified. The result can be increased cost of administration and decreased ease of communication. Also, inexpert classification decisions furnish material for later attack on the entire control program.

Third, the judgment required for proper classification decisions must be made within a frame of reference by which the real risk to the enterprise of improper disclosure can be evaluated. That is actually the concrete application of criteria to the standard. Confused understanding or lack of a sure sense can set much of the program awry and result in general failure to achieve protection objectives.

Because the original classification decision is so important, serious attention must be given to the policy determination of who is to make it. The ideal person is the one who holds the lowest-level *management position*—as distinguished from supervisory or staff professional position—in the subject matter area in which the information is generated. If the classifier is a manager, the required frame of reference will be assured at lowest cost in time or money.

Of course, classification decisions could be referred even higher. A better score of proper decisions might thereby be achieved, but the gain would usually not justify the cost in delay. A manager, by general definition, is one who can take independent action in the best interests of the firm. By limiting managers to original classification decisions in their own areas of expertise, the percentage of accurate determinations of the substantive value of information will be increased. Both a sales manager and a laboratory manager, for example, might have reliable concepts of what would constitute harm to the company, but neither would be able to apply that insight to data in the other's area. That is the reason why managers, although generalists of a kind, are not held accountable for performance outside their assigned areas. The natural result is specially heightened awareness in the assigned area and lessened sensitivity to the unique content of other areas.

It is often suggested that the person who is in the best position to determine the sensitivity of material is its author. That is sometimes but not always true. The author may be competent in the subject matter but lack the organizational perspective needed to apply the protective standard. The cognizant manager, by definition, is both familiar with the subject and qualified to apply the standard. In situations involving authorship by one on a lower level than manager, the author should be required to bring the matter to the manager for determination. The author certainly may assist, but the ultimate responsibility for the decision should be the manager's.

If the manager has the ultimate responsibility, there must be some way to evaluate his performance. The most effective one is to include it as an express item of periodic manager appraisal. That will make clear to each manager that his score in implementing the information control program in his own area will be a factor in his own career growth. The kinds of questions and problems that will grow out of the program and that are discussed in this and other chapters will provide ample opportunity for making appraisal judgments of all managers.

ORIGINAL AND DERIVED CLASSIFICATIONS

The key classification decision is the *original* one. That is the decision made the very first time the data are reviewed—it will usually be contemporaneous with the creation or generation of the information. Once the information is properly classified, it may be reproduced, quoted, abstracted, incorporated, or otherwise assembled and disassembled by many other people who may neither be managers nor be well informed in the subject matter area. However, the later decisions can usually be made much more easily and on the basis of the answer to a simple question: Has the original classified information been reproduced in a recognizable form?

There may be hairline cases in which it is not clear whether all or enough of the key elements of originally classified information have been reproduced, and in those cases recourse to a member of middle or senior management may be necessary. For the most part, however, clerical and supervisory personnel, line workers, and employees in general will be able to apply a derived classification. They will merely continue to label information they process according to the classification it had when they got it. Surely that will require explanation and emphasis in the awareness training efforts, but it need not present a difficult problem. Therefore, it is not necessary to insist that a manager make the decision to treat derived material as proprietary. That decision will already have been made by the manager who was responsible for the original classification.

ELAPSED TIME AS A CLASSIFICATION FACTOR

The life of sensitive information can be diagramed with straight-line segments (Figure 1). A piece of information comes into existence and grows progressively less sensitive through the stages from research to distribution until the need for protection ends. The important fact here is that there is a point in time after which sensitive information no longer requires protection. It may not al-

Figure 1. Life cycle of sensitive information.

Information Generated	Research Stage	Development Stage	Manufacturing Stage	Distribution Stage
Extremely Sensitive	Highly Sensitive	Moderately Sensitive	Nonsensitive	Information Declassified

ways be possible to know when that point will be reached, but there will be many situations in which it will be possible to know.

For example, in the marketing area, if a bid price is highly sensitive because of intensive price competition, it will be so only until a bidder has been selected. After that the reason for protection is gone. The standard no longer applies because disclosure of the bid price after the award has been made cannot result in harm to the company, whether or not the company was the successful bidder. In that case, then, a future event, the award of the contract, determines the end of the period of sensitivity for the piece of information.

In other situations an exact future date after which no protection will be required may be set. In either case, whether a contingent future event or a certain future date is the termination point, it is highly desirable to use the automatic technique for declassifying. It avoids the buildup of unnecessary volumes of protected data without a need for second or subsequent decisions.

There is an important caution in applying the timing technique: it should be reserved for types of data that do not constitute trade secrets. Remember what was said earlier about trade secrets: As long as they continue to meet the legal definition, which, of course, means that they are still not generally known, the proprietor has the legal right to protect against unauthorized use resulting from improper disclosure. That right dies when the information is permitted to enter the public domain. A statement by the proprietor himself that a trade secret of his is not classified after the happening of a future event or after a future date would be a very damaging admission in any litigation alleging the information to be a trade secret.

In any event, the great mass of other data—financial, marketing, organizational—is amenable to selective application of the automatic declassification. The general policy should so provide.

REGULATING DISCLOSURE

So far it has been suggested that the general policy should establish both a standard and a set of criteria by which sensitive information is to be evaluated, should involve the senior executive management in the determination of the criteria, should assign a specific
responsibility for initial or original determinations about specific
items of information to the lowest-tier manager, and should permit
the use of automatic or time-related techniques for declassification
of sensitive data that do not involve trade secrets. The next step is
to regulate the permissible disclosures of sensitive information.

The basic rule governing disclosure is that it will be made only
to persons who require possession, knowledge, or access in order to
accomplish the duties of a position or contract with the company.
Thus, the permissible population to whom authorized disclosure of
proprietary information can be made consists of employees and
nonemployees who are bound by agreement to the company.

It may be urged that there will be times when it is in the company's interest to disclose information to nonemployees who are not
or who refuse to be bound by agreement. One example might be a
supplier of unique materials or services, a "sole source," who is un-
willing to accept sensitive information as such or agree to protect it.
If the company actually has no alternative to such a disclosure (a
rather extreme position that timely planning will help to avoid),
then it should be realized that ability to protect the data will end
with the disclosure. In light of that fact, serious consideration should
be given to declassifying the data to avoid future administrative
costs resulting from internal attempts to continue protection.

It may be that the outsider who receives the unrestricted disclosure will in fact protect the information even though he does not
agree to be bound to do so. If that is so, there is some advantage to
continuing the classified handling, even in the clear knowledge that
disclosure is a major vulnerability that could be actualized at any
time. Decisions of that kind are always hard and serve only to point
up the need for planning to avert the problem. The very existence
of an information control program should stimulate requisite planning once the responsible company function managers see that information that has been classified may later present a disclosure
problem unless procedures are changed before then.

We have already established that disclosure should be limited to
persons with a need to know. The responsibility for deciding who

has that need should be placed on the individual who is to make the disclosure. That is not unreasonable, but it may cause a problem if the relation between the one to disclose and the one to receive the disclosure is not clear. For example, an engineer who has never had any direct dealings with the purchasing department may have to specify an item needed for an R&D project. To give the buyer a clear understanding of the requirements and permit optimum purchasing discretion, the engineer may have to describe functions that are classified. Even though he has never dealt with the individual buyer or perhaps even with the purchasing department on a direct basis, he can apply the disclosure principle with little difficulty. The buyer is required to buy certain items for the company, and the needed item is among them. To buy the item properly, the buyer must know some functional details. They are classified. The buyer has a need to know those details to perform his job. Here is a clear case for authorized disclosure.

In general, need to know may be presumed when the relation between the persons is such that the one who discloses could not continue or complete his work without making the disclosure and the one who receives the disclosure could not begin or correctly perform his work without receiving the disclosure. If in particular situations that relation is not clear to the parties, guidance from the superiors of the one to make the disclosure may be necessary. In any event, the rule should be applied to advance, not retard, the company's interests.

Among the most difficult need-to-know cases are general meetings, across departmental or disciplinary lines, in which free-ranging discussion involves classified items. It should be the duty of the senior representative at such a meeting to assure that the need-to-know determinations required by others in attendance have been properly coordinated before the meeting. With the proper awareness training and high-level-executive monitoring, adequate planning and protective attitudes will become standard.

Disclosures to contractors or other outsiders who have executed protective agreements with the company require closer scrutiny than disclosures to employees. First, it is necessary to assure that there is a binding agreement. That may be ascertained readily only by a more limited number of company personnel than those who will actually deal with the outsider. Second, the relation will *not* determine the appropriate extent of disclosure, but the scope of the contract will. Not every employee who makes a disclosure may have knowledge of the contract scope. It is therefore important that a central clearing authority be established to oversee disclosure to an outside contractor. The policy can provide that the senior executive of the activity having the need for the outsider make the overall determination and instruct other company personnel. Some specific guidance is required beyond the basic need-to-know requirement.

When disclosures are to be made to persons outside the company who are not bound by agreement, including all press and public relations disclosures, a named senior management executive should be responsible for disclosure approval. All planned disclosures should be routed to him first. That does not imply bureaucratic delay when speed is essential. There will always be cases in which an urgent need for disclosure to outsiders arises. There is no reason why verbal or even telephone approval for such a release could not be sought. It is the responsible decision that should be insisted upon, not a ritual for getting it.

WARNING THE RECEIVER OF CLASSIFIED DATA

If the control program is to have continuity, every person to whom a classified disclosure is made must be informed that the information is classified so that he will be cued to discharge his personal obligations in handling it. Classified documentary information is generally so marked, but fully as much information is disclosed orally or through visual access to places and processes as is disclosed by document. The work classifications and identities of the persons who receive nondocumentary disclosures are likely to be different from those of the persons who receive documentary disclosures, but the need to protect is the same.

Every disclosure should be preceded by a statement that the information is classified and at what level. That means that if persons are admitted to areas that contain classified operations, they must be so informed before admittance. It also means that persons who are to participate in a discussion of classified items must be so informed beforehand. It is regrettable that nondocumentary disclosures are often made without notice and that insufficient attention has been paid to them in the information control programs of many firms. Particular care should be taken that training routines and materials are developed to give this program element proper emphasis.

PREVENTING UNAUTHORIZED DISCLOSURE

Later chapters will deal in detail with the use of protective routines. It is important to note here that the basic policy should provide for such routines, at least in general terms. What can readily be made standard throughout the entire enterprise should be standardized in the policy directive. That will very often include overall requirements for the marking, transmission, storage, and disposition of classified documents, the controls over physical access to classified places and operations, the requirement for the execution of agreements and acknowledgments by employees and others, and the need for briefing and debriefing persons who have access to classified company information. The Appendix shows a model policy statement that contains illustrations of provisions for meeting all those requirements. It might be reorganized to suit the format and other requirements of a given company, but it can be used as a checklist or planning guide.

REFERENCES

1. Executive Order 10501, November 5, 1953; Department of Defense Directive 5220.22; Department of Defense Manual 5220.22M, as amended.
2. Report of the Commission on Government Security, published pursuant to P-L 84–304, June 1957.

8

---◆---

THE COMPUTER
AND ESPIONAGE

Computer technology has opened whole new areas of security vulnerability because of the simultaneous operation of three factors. First, data that formerly would have been dispersed into many different locales are now brought together—sometimes kept together —in the computer. Second, to optimize remote operation of industrial enterprises, computer facilities are being widely designed for remote-terminal access. That has the effect of exposing the information newly concentrated in the computer center to persons outside the center, perhaps thousands of miles away. The only requirement in many cases is access to a suitable terminal, or perhaps merely to a telephone instrument. Third, the economies inherent in broad computer use by small enterprises has created the computer service bureau.

Service bureaus are commercial computer installations that rent machine time to a variety of users. In some cases the user brings his data to the center for processing; in others, the user gains access to the computer remotely by means of some form of telecommunication link such as the data telephone. In a very large number of service bureau situations, many customers use a single computer for many different programs and each of them gains access in accordance with an established schedule that divides the available machine time among them.

It is in cases of multiple use that the problem of data exposure reaches its compound dimension. Data are concentrated at the user's own location and then transmitted to the service bureau. At the bureau they are loaded into the computer and allied devices and remain available there not only for the authorized customer but perhaps also for unauthorized persons who have found the key into that particular system. While at the bureau the data are also often concentrated in some conventional hard-copy format such as a printout or management report. They may be vulnerable to persons at the center who are authorized to be there but who intend espionage at an opportune time. That, of course, could be true of any member of the bureau staff, about whose selection or reliability the user most often has little or nothing to say. The data may also be vulnerable to unauthorized persons who, through inadequate security, gain access to the center. And they may be vulnerable to the electronic trespasser.

The perils of information loss due to one or all of the cited security vulnerabilities are really so great that many managers have chosen, perhaps unconsciously, to ignore them. "Ignore" does not mean to take absolutely no precautions, although there are cases in which that is literally true; it means to rely on an established security ritual that is not even responsive to the nature of the threats. For example, consider a company that has a completely dedicated, captive computer system involving remote terminals in all its offices and factories. It may protect the home office computer installation with access control techniques and need-to-know theory, and it may even take certain measures to harden the computer site physically. But if it does not take measures to control the remote terminal activity, *both* as to the persons authorized to gain access to it or process data from it *and* as to the nature and extent of such data, the home office routine may be illusory. A determined information seeker would attack the weak point, the terminal, and not even bother with the center.

Now consider the user of a service bureau. The data involved may be treated with extreme care while in the user's own possession, and may even be transmitted to the bureau under security control. While in the bureau's hands, however, they may be exposed to technicians who are not screened prior to employment, who have not been adequately prepared by security training, or who, in the interests of the service bureau's economy of operation, are granted broader access to customer information than the customer itself would permit. Such exposure can result in selective or wholesale copying of customer data without the customer even knowing to whom its information has actually been exposed.

Ethical service bureaus are aware of the problems stated and are making serious efforts to reduce exposure. However, a commercial firm is in business to generate profit. There are limits beyond which no further investment is justified because it would have an unfavorable impact on profit. No single firm will voluntarily introduce costs that it must absorb or that will make it noncompetitive. In such cases, lacking insistence on the part of users, minimal security precautions that are not really effective may be the only barrier to major compromise. The victim, if loss occurs, is the ultimate user, not the service bureau. It is therefore up to the ultimate user to be satisfied that adequate security precautions have been taken or to calculate that information is likely to be compromised.

If users of service bureau facilities are to make informed judgments about the security of their information, they must be familiar with the equipment and operations in use at the contract bureau. For that reason, even the relatively small business that uses outside bureau computers and plans no investment in its own main-frame facilities cannot avoid the need to understand the system for which it contracts. Whether a firm owns or rents computer equipment for use on its own premises or purchases machine time at another location, it cannot do an adequate job of protecting its information until it understands the equipment in use and becomes familiar with the process controls in effect at the site. Loss of information from computers is effected both by the internal security weakness of the equipment and its related software and by the operating conditions at the equipment site.

THE EXTENT OF VULNERABILITY

To establish workable protective routines for computer-generated or -processed information, it is first necessary to understand the full extent of exposure. The location of and medium for each sensitive datum must be known. Because the computer center may take input from users who do not process the information to any form beyond hard, clear text copy, it is important to follow the flow of such information all the way from the ultimate source to the computer. In the same way, because the computer will generate new or re-arranged data that may appear in clear text in the form of management reports and other printed output, it is necessary to trace all computer output to the ultimate user.

Intermediate-status data—that retained in the computer, in peripheral devices that are part of the computer installation, or in physical storage in the custody of the computer center—must also be accounted for, particularly when access to it is possible by computer center personnel who are working on other assignments. The flow of data is from all ultimate sources or inputs, through the data center, and out to all ultimate users. Adequacy of security must be achieved at every stop on the route.

INPUTS TO THE COMPUTER

We will first examine the physical inputs under the overall control of the computer facility owner. Later, the more complicated electronic inputs through terminal devices and delivery to an outside service bureau or other computer utility operator will be considered.

Data will come to the captive or in-house computer center in three forms: (1) plain-text alphanumeric information as in sales reports, accounts receivable and payable vouchers, and other conventional documents to be converted to computer format, (2) intermediate or partly processed data that have been translated into digital character as in punch cards or punch tape, and (3) fully machine-ready data as on magnetic drums, disks, and tapes.

The second category, or semiprocessed data, involves a preparatory activity that often consists of a unit-records or key-punch operation in which the input information is transferred by key-punch machines from alphanumeric clear text to some form of coded representation. The presence of a punched hole in a particular location (field) on a data card has a specific meaning. By variously combining punched holes, the letters of the English (or other) alphabet, the digits 0 through 9, and such symbols as the asterisk, dollar sign, comma, and period can be entered on the card. Obviously, anyone with a knowledge of the card design and the punching code can read the information by simply performing in his mind the usual machine function of replacing the punched hole combination by the letter, numeral, or symbol for which it stands.

The design of unit-records data card content configuration is treated extensively in many publications and will not be detailed here. Most practiced technicians who work with such cards can read them easily. The point here is that the information, while nominally encoded in digital form, is actually still so close to clear plain text that, even without a sorter, printer, or other mechanical device, a practiced person could read the cards and extract their information content. Information in this medium is thus highly vulnerable to compromise.

Although some plain-text input data are processed at the unit-records facility into punched cards, a similar input medium, punched paper tape, can be generated directly by certain office, store, and factory equipment. Cash registers, for example, often capture the transaction data (amount, product code, tax, date, position, cashier, and so on) on continuous paper tape contained in the register itself. Periodically the tape is unloaded and transferred to paper-tape-processing units in which the data are further processed for input to the computer.

The coding on paper tape is similar to but not exactly the same as that on cards. Tape is much denser in respect to the amount of data contained, and some entry point from which visual examination should begin is required to correlate the punches observed with actual transactions. A practiced person reading a data card would know the transaction limits from the card itself; he would have no problem of where to begin and end. Were he reading tape, he would need the beginning point to recognize distinct transactions. Therefore, it is a little harder, but not impossible, to retrieve plain-text information by visual examination of paper tape.

To recapitulate, the exposure to data compromise, when computer input consists of punched cards or tape, lies in visual access to the medium and the susceptibility of the medium to copying. Photocopying ordinary documents is a common technique of information theft. Punched cards and tape also can be photocopied: access to the proper printing machinery will permit the punched cards and tape to be reduced to printed hard copy. Furthermore, access to appropriate other unit-records equipment will make possible the cutting of one or more duplicate decks of cards or rolls of tape from the deck or roll in possession. Blank data cards and paper tape are given minimal accountability attention in most facilities because of their very low cost. The use of several hundred, or thousand, cards or of one or more paper tapes to make unauthorized copies would probably never even be noticed. Obviously, the possession of the coded cards or tapes plus access to blank material and the appropriate machines on the premises and time in which to remove the cards or tapes to the machines for copying and then return them are all that are needed to compromise the information.

Another form of input that is very close to plain text is the optically read or magnetically inked one. Examples are the typical commercial credit card, with its embossed symbols, and the preprinted bank check, with its square-block numbering. In both cases the character is clearly perceptible to the unaided eye with a little practice. Look at the customer copy of any credit card transaction. You can read the data imprinted by the embossed characters. Look at your checkbook. You can read the numbers on the checks. Any input medium using optical character recognition (OCR) is likely to be even easier to interpret than punched cards or tape. An exception would be OCR in which the character itself is coded and a key is needed to relate the coded optical character to the alphanumeric value it represents. Most OCR, however, uses some readily identifiable form of conventional numerals and letters.

MAGNETICALLY ENCODED DATA

The last step before information is ready for actual processing by computer is information conversion into magnetic bits (*binary digits*). Electronic processing impresses magnetic charges on sections of the medium—the drum, tape, or disk. Since the information is in binary or two-state form, the presence-absence or positive-negative alternations on the magnetic medium are the basis for recording all information. Unlike punched or OCR media, magnetic encoding is not readable visually. The medium must be processed through a read-write head, which will either sense magnetic coding already on the medium (read) or code new information on it (write). The results of a read cycle will be processed further and either transferred to another machine location, such as the core memory, copied on a duplicate medium, or converted through the printing process into plain-text hard copy.

The vulnerabilities of sensitive information in magnetically encoded binary digital storage are to unauthorized copying, compromise through conversion of the data to plain text, and, sometimes overlooked, simple theft of the medium itself. When the medium is as small and easily transported as a magnetic tape, the physical theft possibility becomes a probability.

THE KINDS OF DATA INVOLVED

When computer operations are studied, two general types of data must be distinguished: (1) the master or base data—the pieces of information that are to be processed, and (2) the program data, the instructions to the computer on how to process particular base data. Frequently, both kinds of information are sensitive. Base data can be sensitive because they are merely sensitive information in a different format, information that is discussed in Chapter 2 in detail. Program data can be sensitive either because they represent a distinct business advantage in and of themselves or because they indicate how the computer can be manipulated to gain unauthorized access to base data being processed.

A case involving two computer facilities that operated as utilities or service bureaus illustrates both aspects.[1] One service bureau had developed a proprietary program for handling client data that it regarded as confidential and highly valuable. It was, in fact, selling its services to certain clients on the basis of that very program, which allowed it to optimize its time-sharing schedules. An employee of a competitive computer utility that performed similar work and shared at least one client was aware that access to the first bureau's computer, operating in a time-sharing mode, could be gained by telephone by using an unlisted number normally given by the first bureau only to its regular clients. Further, once communication between user and computer was established, access to program and base data could be achieved by the user's keying in an assigned user code.

The first bureau has charged in a legal action that, with knowledge of both the unlisted telephone number and an authorized client's use code, the competitive bureau employee was able to gain access to the computer and the target program and to retrieve certain key information for unauthorized use. Presumably, the objective was to steal the first bureau's proprietary program (the base data were unimportant to the acquisitive second bureau) and use it to improve the competitive bureau's own time-sharing operations. The case illustrates that programming information can be valuable both as a key to the computer (the dial-up, user-code-entry technique) and for its own sake (the trade secret aspect of the proprietary program being sought).

DATA IN PROCESS

So far, data on the way to the computer have been discussed, and now data in process must be considered. Data in process in a computer as a captive facility are vulnerable to compromise in two ways: (1) They can be duplicated during normal operations either by a change in program instructions or by computer operations personnel making unauthorized use of facilities at unauthorized times. (2) They can be intercepted from outside the computer center. Clandestine interception can be either by attack on the electromagnetic envelope surrounding the computer or by coupling in some way to the telecommunication links that bind several computer units or their users together.

ELECTROMAGNETIC RADIATIONS

Computers utilize electromagnetic energy in the high end of the radio-frequency spectrum. In addition to the normal, directed transmission of data from device to device along wire or microwave or laser connectors, random radiations are created near the signal-generating equipment and along the transmission lines utilized to pass the high-frequency signals. It is possible, given the correct interception and demodulation equipment and close enough proximity to the target, to capture the random radiations, demodulate them (convert them from high-frequency pulses to plain text language), and leave no evidence of having done so. The espionage agent who practices that technique could record the captured data for later processing at his leisure. That would mean a minimum of equipment to be brought to the computer site for the interception. The controls on this type of exposure are a combination of widening the distance between the computer installation and uncontrolled neighboring premises and screening the computer electronically. Both controls are discussed later.

The second method of attack is to couple to the telecommunications channels, which are special cables or even standard telephone pairs. The attack can be by a classic application of the tap technique: physically connecting to the target cable either along its course or at a switch junction such as a terminal board, a private branch exchange (PBX), or even a telephone utility company switching facility. All that is needed is knowledge of the cable or pair location and access to the location long enough to make the connection.

The attack can also be by inductive couple, in which there is no physical connection to or penetration of the target link. The couple features the use of an induction coil that is so simple that a workable model could be made by any high-school student studying elementary physics. (The topic is discussed more fully in Chapter 6.) The Omnibus Federal Crime Statute [2] has made the use of either such devices and their more sophisticated relatives or the information obtained with them a lot more dangerous for the espionage agent. That will surely deter some but cannot reasonably be expected to prevent all industrial espionage. Active defense is needed when the target is sufficiently valuable to provoke attack interest.

DEFENSES FOR THE CAPTIVE FACILITY

The defenses for computer input data entering the captive facility are basically the combination of need to know, accountability, access control, and physical safeguards discussed in other chapters. The first step is to analyze *all* the input data coming from *all* the sources and to earmark those that are sensitive by their very nature. That involves the application of information classification described in Chapter 7. The next step is to assure that the input data that are sensitive are protected by disclosure, transmission, storage, accountability, and use safeguards of the kinds described in Chapter 9. The application of those same techniques, incidentally, will be appropriate for the protection of data leaving the computer center in conventional, hard-copy form. The specific defenses outlined in the rest of this chapter deal with the unique vulnerabilities of the computer and information being stored in or processed by it.

ADMITTANCE TO THE COMPUTER AREA

1. Every location of either operational computing or peripheral equipment (printers, disk drives, and so on) should be under positive entrance control, and only those persons whose work assignment actually requires them to work with such equipment should be regularly admitted. Generally that means restricting regular ad-

mittance to prescreened computer operations personnel assigned to work the particular day and shift.

2. Persons who require occasional admittance or admittance under extraordinary circumstances (service and repair personnel, users of the computer facility, porters, and the like) should be under visual surveillance. Their admittance should be recorded by date and hour, and their movements and activities within the computer area should be only those necessary to discharge the purpose of the visit. Conscious observation by normally assigned operations personnel can achieve such restriction.

3. Allied departments such as systems analysis, programming, and unit-records operations present a special admittance problem. There is a natural tendency to regard them as an integral part of the computer facility and to permit free access. That is particularly true of programmers who are either designing or modifying programs. The practical value of being near the computer lies in the speed with which transactions can be made. However, one of the greatest problems in computer security control is access to the computer by persons who are familiar with the computer's theoretical and mechanical operation and have access to base or program data. Not only compromise and theft of information but also a wide variety of frauds have been accomplished by people with such access.[3]

4. During hours when the computer and related areas are not operational, positive controls should be maintained to prevent or at least detect entry. Intrusion alarms with reasonably prompt response capability and reliable locking devices are needed.

INFORMATION STORAGE MEDIA

5. Program and base data will be stored in magnetic format on disks, tapes, cards, and drums and, of course, in the main-frame memory. Data will also be stored on punched cards and tape and in plain-text hard copy. In the latter class will be documentation files maintained by programmers regarding specific program development and maintenance. Such program documentation may be found either at the computer center or elsewhere, as in the programming area. The other stored information will normally be in or closely adjacent to the center and will constitute the library.

Each sensitive item in the library should be positively identified. That means externally marking disk packs, tape reels, card magazines, and drums in some distinctive fashion. The marking will permit immediate visual recognition of the sensitive item. In addition to the visual identification, the recorded data file is always internally identified by its name and file number. That information is magnetically encoded or punched in header or locator positions. The sensitivity level of the file should be included in the header so that, when the file is actually being run on the computer or peripheral device, it will be amenable to special security instruction and control. Sometimes such controls are contained in supplementary programs that are specially prepared to execute security procedures before and during use of the file data. The internal file header will also contain a reference to such supervisory or supplementary routines to assure that the security measures are taken.

A good illustration of the dangers in unauthorized access to both data and computer can be given here. Assume that a file is sensitive enough to require supplementary special security programming. Now, if the main program file can be modified by deleting its internal reference to the special security program, it can be run without the security controls. That might permit later use or copying of its data at a time of convenience by an unauthorized but knowledgeable person. After his unauthorized use, the person could replace the internal reference in the main program and leave no specific evidence of the modification or unauthorized use.

6. Sensitive base and program data should be stored in such a way that their withdrawal is always on an accountable basis and the date, time, reason, and user are recorded. The storage can be supervised by a librarian and controlled by a written charge record or by a semiautomated one that involves an identification device. In the semiautomated system an authorized user has identification that is capable of making a permanent record. For example, a credit-type plastic or a similar card with some distinctive coded information is inserted in a reader-recorder at the time the file is withdrawn. If the recorder is coupled with a real-time clocking device, the identity of the person to be charged with the file and the date and time of withdrawal are recorded independently of the librarian. The librarian, however, verifies the identity and issues the file. In that way the librarian alone could not issue a file to one person and record a different person's identity. Moreover, no file could be issued to any user who was not in possession of his identification device unless by way of an exception routine involving higher authority in the enterprise.

7. Sensitive files removed from storage should go only to the computer or peripheral equipment area or to an approved alternate location such as a security vault. Removal of a sensitive program or base data file to any other location, including a user area, should be by way of an exception routine involving higher managerial authority and a documented record.

8. Transmission of sensitive files from the library to the computer or security storage area should be secure enough to prevent surreptitious substitution of one file for another en route. That is a good security reason for making the storage library communicate directly with the computer. Operational convenience, of course, suggests the same thing.

If sensitive files must leave the contiguous physical areas of library and computer, they should be locked in suitable containers and be in the custody of an authorized person. It may happen that ease of use and convenience are reduced by the secure transit precaution. However, the balance to be struck is between the harm to the enterprise that is potential in unauthorized disclosure of the information being moved and the penalty in reduced convenience. If the unsecured movement offers real opportunity for substitution or copying and the information is truly valuable by the standards suggested in Chapter 7, then the self-interest of the enterprise requires the protective routine to be designed into normal operating procedures.

9. Access to the storage area should be restricted to persons who are responsible for the storage activity. If others, such as porters, utility personnel, and maintenance crews, must be admitted for special purposes, they should enter only under the visual surveillance of someone who is familiar with the stored files, and their activities while present should be closely limited to the stated purpose of the admittance. If sensitive files are kept in specially locked containers or are otherwise effectively segregated within the storage area, control of casual admittance, though still necessary, will not be as cumbersome.

Situations that seem to justify exceptions to the restricted storage area access will arise. For example, night or weekend operations may be conducted with a skeleton crew. (Sometimes that means a single operator working alone, and that is not a sound practice for safety as well as security reasons.) Breakdowns or unforeseen difficulties may require that the work scheduled for that shift be abandoned or delayed. To avoid complete loss of the shift, substitute programs must be available. The argument, then, would be to permit the operator(s) to have access to the storage area to assure continuance of operations. And the answer is that that is precisely the situation to be avoided. The operator(s) could gain access to the library, withdraw whatever sensitive programs or data files were available, copy or run them, return them to the library, and still pull sufficient contingency work to cover the interrupted shift.

At the very least, some authorized supervisory member of the computer operations group should be involved in library access. If a supervisor is present during the shift, then all file withdrawal should be his personal responsibility and the locked file storage area should be accessible only to him. If no supervisor is present, then the earlier suggestion that sensitive files be specially segregated may be the only economical way to protect proprietary data and optimize computer operations. In that event the contingency work done to fill in for interruptions in the normal scheduling must not include sensitive files. If the volume of sensitive files or the nature of operations is such that access to proprietary data cannot be denied, the practice of unsupervised library access must be reexamined. It is not justified in the presence of high-probability loss of high-value data.

To assume, however, that merely because a supervisor is involved in the file access routine the whole problem of unauthorized exposure is solved is to be extremely naive. There is convincing evidence that more extensive and costly frauds are perpetrated by supervisory and managerial personnel than by other classes of worker. By requiring *more than one person at different levels in the organization* to cooperate in the withdrawal and use of sensitive information, the probability that unauthorized use will occur is dramatically reduced. Now withdrawal requires a conspiracy in which each participant implicates himself. Although conspiracies occur often enough, more is risked in them than by one person operating secretly. The conspiracy probability is further reduced by each additional person in the transaction.

10. When sensitive data on a storage medium are no longer required and the medium is to be reused, care must be taken that the data are obliterated or made unintelligible before the medium is released for uncontrolled handling. Two procedures are available if the medium is magnetic tape. The data can be erased, that is, the specific character of the positive-negative magnetic coding that comprises the binary digits can be removed by degaussing, or im-

posing a uniform magnetic configuration on the medium. Alternatively, the medium can be overprinted with new or random magnetic coding. The Department of Defense suggests to its contractors that overprinting should be accomplished by making three verified passes over every bit location while using at least as great a signal strength as was used to write the data to be overprinted.[4] That is a good guide for any computer user who is desirous of removing previously sensitive magnetic data. One overprint would probably suffice in most cases because the resulting garble, even if each original bit were not completely masked, would render recovery of useful data unlikely. For the most sensitive data or for the release of the storage medium outside the enterprise after purging, either degaussing or multiple overprinting is recommended.

To desensitize punch cards or paper tape, the lacing technique by which all data positions are punched is appropriate. The alternative is destruction by shredding or tearing to a degree sufficient to prevent reassembly of the card or tape. Ultimate destruction of sensitive-medium waste should be by burning.

THE COMPUTER IN OPERATION

Still assuming a completely contained computer facility with no remote-terminal access, the following precautions are appropriate to minimize information theft or compromise during actual operations.

11. When sensitive programs are being loaded, run, or unloaded, area access should be denied to otherwise authorized visitors unless their presence is necessary for successful operation of the sensitive program involved.

12. All peripheral devices, such as flexowriters and printers, should be cleared of all hard-copy printouts and carbons before the sensitive program precautions are terminated.

13. The main memory, auxiliary memory units, and all temporary storage units not needed for permanent storage of program or base data must be purged of sensitive data at the end of the program and verified as nonsensitive by a complete memory dump. That can be done by incorporating a special security routine in the software for each sensitive program. Alternatively, the routine can be incorporated in the supervisory monitor and activated on cue for all sensitive programs. The cue could be a datum in the header and trailer sections of the individual sensitive programs. The specific

approach is dictated by the kind of hardware and software generally involved. The in-house or utility bureau program staff can provide the necessary routines as part of the original program design. What is critical is to flag the requirement for a security purge in the very first outline flow chart of the sensitive program.

14. Some or all of the programs being run or planned may be sensitive enough to warrant electronic screening of the computer center to prevent capture of electromagnetic radiations. Screening involves the installation, around the entire periphery of the center, of metallic mesh to intercept and ground out any radiations. Screening is expensive and must be done with precision to assure that all radiations are trapped. Cables and other leads that go beyond the screen must be filtered at the point of exit from the center to remove their radiations. Normally the cost of screening should be considered at the time of construction of the computer site, because the installation can cost considerably less if it is done during rather than following completion of construction. Any computer center that cannot reliably assure significant spatial separation from public access or other occupancies and that is running highly proprietary programs should periodically consider adding screening when the possible cost of loss of information is notably greater than the cost of installation.

15. As previously noted, telecommunication connections, such as telephone, Telex, and TWX, that enter the computer area are targets for inductive or physical tapping and should therefore be checked frequently by competent personnel. The check should include the instrument and the wire leads from the instrument to the first terminal board or other switch connector where the leads are combined in a common cable. From that point on the masking effect of communications on the other leads or pairs in the cable will probably protect the pair containing computer center communications. While the computer center pairs are isolated, however, they are very vulnerable. (The whole problem of communications interception is treated more fully in Chapter 6.)

16. In a well-run computer operation all jobs are scheduled and correlated with the withdrawal of storage media such as disk packs and tapes. It is important that responsible operations supervisors regularly check the storage media in the center against schedule sheets to be certain that unauthorized material has not been included to await an opportune moment to be loaded and run or copied.

17. All machine time should be accounted for. The required length of a given sensitive program in elapsed running time will be known. The machine time actually clocked for that program should be logged. Variances in the machine time from the established time should be *invariably* and *completely* explained in signed and certified log entries that explain the error or exception that caused the variance. The error logs should be reviewed by each succeeding shift supervisor and daily by senior operations management. Extra machine time required for exceptions during the running of highly sensitive programs should be regarded with suspicion and regularly verified.

18. As part of the emergency or disaster plan prepared for the computer center, the situation in which sensitive programs are running or exposed at the time of an emergency should be considered. If evacuation of the computer room is required, someone should be able to certify that all personnel actually leave. Normal access control routines should then prevent reentry except under controlled conditions. The sensitive exposed storage devices should be inventoried as one of the first recovery steps after the emergency. If they are removed as part of the emergency response, then the rally point or secondary control location to which they are taken should be clearly identified in the plan. The material should be inventoried and secured immediately upon its arrival there.

THE REMOTE-TERMINAL ACCESS PROBLEM

In addition to all the problems already noted, any computer to which access is gained from remote-terminal devices is vulnerable to unauthorized compromise of data by skillful use of those terminal devices. The problem is at its worst when the computer is operating in a multiprogram, multiprocess mode and not all the remote users are part of the same organization. The service bureau whose clients run sensitive programs is a good example. However, as current users of major computer installations go more deeply into real-time applications and extend the area of remote terminals to international dimensions, even single-enterprise systems will be badly exposed.

Data losses due to remote-access vulnerabilities can occur in three ways: (1) An unauthorized user can learn an authorized user's key and program instructions and deliberately retrieve sensitive information. (2) Through an error or malfunction within the system, sensitive data can be displayed or delivered to unauthorized terminals. (3) Unauthorized persons can modify program or base data during remote access to permit future unauthorized retrieval or disguise prior manipulation.

The countermeasures are clear:

1. Only authorized persons should have physical access to a remote terminal during sensitive operations.

2. Only authorized terminals should be able to go on line to sensitive programs and only after *both* the authorization of the terminal and operator *and* the secure status of the terminal environment have been verified.

3. Departures from program parameters in sensitive programs should be closely monitored, and such events as improper query, query out of sequence, nonstandard input, and nonstandard response should result in one or more penalties of the following kinds: The program is discontinued until appropriate coded reentry signal is received. The computer center logs and prints the date, time, and nature of the irregularity. An alarm is signaled at the computer center or other location from which qualified response can be made promptly.

Countermeasures are easy to itemize, but their implementation may be quite complex. The nature and range of purposes of the remote-terminal hardware will determine in very large measure just how serious the problem is in a given situation. For example, a terminal that can only retrieve and read cannot make program or data changes. A terminal that operates in only one language (say formula translation—FORTRAN—for scientific problem solving) will not be able to interface with a main computer monitor or operating system written in another language and manipulate other sensitive data. A terminal that can query, update, and delete, however, can attack base and program data unless great care is taken in imposing hardware and software lockouts on the terminal.

In a limited treatment of such a comprehensive topic only the highlights of corrective routines can be noted. Readers are encouraged to consult one of the specialized works now available on computer security for more comprehensive discussions.[5] As a general checklist, the following countermeasures should be considered:

1. The use of lock words or key words to authenticate a terminal before allowing it on line with sensitive data. Lock words can be as complex and changed as frequently as conditions warrant. Manifestly, they should not be the logical first guesses of an attacker. In

that class would be the usual operator's name or the abbreviated name of the office in which the terminal is located.

2. Second and higher levels of authentication can be required for successive levels or increments of a given program. Sign-countersign arrangements in which both the sequence and the composition are critical can be designed.

3. Bounds registers should be provided in the memories so that access cannot be gained to machine addresses that do not contain data relevant to the terminal-selected program.

4. Absolute machine addresses, that is, main-memory locations at which given items of data may be stored at any moment of operation, should not be accessible to the remote terminal. Relative addresses are adequate. That means that the terminal has a programmed address to or from which to process data. Machine operations may move the address to another location in the memory. If they do, the supervisory program will provide a mediator or formula by which the programmed address, if input from a remote terminal, will be converted to the actual address. The formula or mediator itself should not be available to the terminal.

5. Queries from the remote terminal can be programmed to start a series of events in which the on-line connection is first broken and then reestablished *from the computer, not from the terminal.* That would prevent an unauthorized person from simulating a system terminal and thereby gaining access to base or program data. Of course, the added communication expense should be calculated in the overall cost-benefit model of the security program. In sufficiently sensitive cases, however, the added cost may be the least expensive way to assure the identity of the remote terminal.

6. A query from a remote terminal may also be the basis for collateral verification of the terminal's identity. For example, a terminal input could trigger a separate telephone inquiry to the remote-terminal voice-grade telephone number listed in the computer center. Voice-grade contact with the terminal having been established, a validation routine involving the computer, the remote terminal, and the telephone could be worked out. Another approach, particularly in situations in which the computer or the subscriber has low unit message cost communications such as wide-area telephone service (WATS) available, would be to verify the remote terminal's identity by a telephone call to a number that is located elsewhere than at the remote terminal but has low-cost voice connection with the terminal and the central computer.

Since this book is concerned with espionage—the intentional effort to obtain unauthorized data—extensive comment is not made upon protection of computer facilities from physical security threats. However, the very serious vulnerability to innocent or malicious (sabotage) error introduction does suggest that the accuracy of the data transmissions be monitored too. All the usual techniques of parity checks, multiple transmissions, hash, and columnar and cross-footing totals should be considered as particularly appropriate for transmission between the central computer and the remote terminal.

CRYPTO SYSTEMS

Ultimately, the sensitivity of data involving remote terminals may require the use of encrypting and decrypting techniques to convert the data transmitted to an undecipherable state except after processing by a key. The computer can be used to design its own crypto system, but that would expose the system to compromise by anyone granted access to the machine-programming schemes. To the extent that the reliability of computer personnel at both the central computer and the remote terminals can be assumed, computer-designed crypto transmissions would at least protect the data in transmission from surreptitious interception.

Crypto keying devices that are separate from the computer and unavailable to computer center or terminal operators are the best solution at the present time. With them, the encrypting is done by the keying device, which is applied to the data stream that leaves the computer. It is then applied to the data stream received at the remote terminal to decrypt data before they are displayed or printed out. A crypto keying device is required (and is normally not easily procured or designed), and special personnel must work with the crypto key but be distinct from computer operations personnel.

In continuously sensitive operations the encrypting-decrypting approach would require a parallel organization (small, perhaps) of persons who were qualified to encrypt and decrpyt and who would not permit the terminal or computer operations personnel access to the crypto keying device or to information about its operation. At the present time that approach is used almost exclusively by the military and intelligence agencies.[6] However, it is highly probable that, within five years or less, commercial crypto systems will be available. Until then, the organization with serious espionage vulnerabilities in its computer operations should be certain that it employs or consults highly qualified systems design and program design experts who are familiar with data protection design solutions and that it takes maximum advantage of the computer manufacturer's protective architecture.

The prospects are that the protective architecture will be improved by an IBM study, to cost $40 million over five years, of "techniques for assuring the confidentiality of data stored in computers." [7] The study will be conducted at customer as well as IBM facilities because, as IBM's chairman observed, the features of the security system can be worked out only in an actual working situation.

REFERENCES

1. *Information Systems Design, Inc.* v. *University Computing Company,* as described by William Godbout in *Security World,* May 1971, p. 22 et seq.
2. 18 U.S.C. 2510 to 2512.
3. See the case discussions in Harvey Gellman, "Using the Computer to Steal," *Computers and Automation,* April 1971.
4. *Department of Defense Industrial Security Manual* (DOD 5220.22M), par. 19h, as revised.
5. William F. Brown (Ed.), *Computer and Software Security* (New York: AMR International Inc., 1971). Leonard I. Krauss, *SAFE, Security Audit Field Evaluation* (East Brunswick, N.J.: Firebrand, Krauss & Co., Inc., 1972).
6. *Cryptographic Supplement to the Industrial Security Manual* (cf. note 4, supra).
7. *The Wall Street Journal,* May 18, 1972.

9

---•·•---

TECHNIQUES OF PROTECTION

Dr. Samuel Johnson is reputed to have said that imitation is the sincerest form of flattery—to which might be added, but not when trade secrets are involved. The discussion in preceding chapters has consistently emphasized that the key to an effective industrial espionage protection program is the prevention of imitation by competitors. Also, it is clear that the courts in the United States have held that two basic requirements must be met if legal protection for secret data is to be expected. First, if secret information is to be disclosed to another party, notice must be given that it is secret and that there is a prohibition against its further disclosure. Second, the secret data must be capable of protection and the owner must take positive steps to protect and preserve their integrity. Specific ways to meet those two requirements so that the hazard of industrial espionage can be minimized will be discussed in this chapter.

Applicant Screening

The first step in designing an effective industrial espionage prevention program is to develop an applicant screening process.[1] A number of the cases cited in preceding chapters, as well as other recorded instances of industrial espionage, clearly indicate that a major hazard is the individual who first gains access to sensitive information as an employee and later either offers the information to a competitor or goes to work for a competing organization and uses the information to the detriment of his former employer. If effective screening techniques have not been established, the most elaborate program of internal controls after the individual starts to work may be useless. Mistakes in the selection of employees will allow the individual who intends to obtain valuable company data to penetrate the organization's first line of defense against industrial espionage: the applicant screening process.

The application form is a basic tool in the screening process and is usually the first and most important document the applicant will submit to the organization. It must be so designed that the applicant is required to give complete information about himself as well as his background and experience. Also, the form should be so designed that the applicant cannot easily omit or misstate information.

After the form is completed, it should be carefully reviewed not only to insure it is complete but also to detect danger signs of a problem, such as job hopping, that might require further development. An interview by an experienced company representative based on the completed application form will often be effective in uncovering data that might cause the enterprise to question the individual's trustworthiness to handle and protect sensitive information.

Unfortunately, it cannot be assumed that each applicant is what he claims to be. Applicants at all levels in every type of organization may falsify qualifications and background data. The press attention given the book about Howard Hughes written by Clifford Irving, who was called "con man of the year" by *Time*,[2] overshadowed another case involving falsification. Donald Angus Stuart, a deputy medical examiner of Los Angeles, was arrested in February 1972 and charged with one count of perjury and two misdemeanor violations of the California State Business and Professions Code. He was charged with faking his medical degree and physician's license. He had been on the Los Angeles coroner's staff for three and a half years.

Apparently an adequate background check was not made before Stuart was hired. An investigation was started only when he failed to apply for a California medical license after he was told to do so. Indications were that Stuart, a naturalized U.S. citizen born in England, claimed when he filed his job application that he was licensed to practice in Illinois and had attended the University of London from 1939 to 1946. He reportedly indicated that he had first received a law degree from the University of London and then a medical degree from the University's Middlesex Hospital School. The investigation is reported to have developed that officials of the University of London had no record of Stuart ever attending classes there, that the Illinois license number listed by him had been issued to another individual, a Dr. Lowell, on September 8, 1914, and that Stuart had been posing as a physician for 25 years.

The obvious question that might be asked is, "How could such an incident occur when one letter to the University of London would have unmasked the imposter?" The simple answer is either that the information submitted in this instance was accepted and no checks were made or that, if there was one, the investigation was not properly conducted. The case is not unusual. Many instances of such falsifications of records have been uncovered by background checks. Management representatives of organizations that do not have effective applicant screening programs would no doubt be astounded at the discrepancies in the recorded backgrounds of trusted employees that could be revealed by effective background checks.

The next step in the applicant screening process is the applicant investigation, which can be conducted by company personnel, an outside agency, or a combination of the two. Regardless of how it is made, the investigation should insure that the data supplied by the applicant are accurate and include all of the important information concerning background that will reflect on the applicant's reliability and trustworthiness to handle sensitive data.

Finally, after all the data are collected, they should be reviewed by a responsible management representative. If the screening process is adequate and has been followed, the hiring decision should be clearly evident to the individual who is responsible for making it. Admittedly, the screening process can never be completely accurate, but at least it should eliminate the applicant who is obviously untrustworthy and who might engage in industrial espionage because of previous activities.

Employees who will have access to secret data should, as a part of the employment processing, be required to execute secrecy agreements before they begin work. Because those agreements are such an important part of an information protection program, a more detailed discussion of them is included in the following chapter.

Education

Employee education is an essential element of any industrial espionage prevention program. The education process should be continuous and be directed at employees at all levels. Ideally, every employee should be convinced that his job, his success, and his growth in the organization depend on the success of the enterprise and, further, that the success of the enterprise depends in large measure on the safeguarding of company secrets. As was pointed out earlier, the basis for the entire prevention program should be a general policy statement that is given wide dissemination. It is essential that all employees be completely familiar with the position of the organization and that they know how they fit into the espionage prevention program and what they are expected to do to safeguard information.

Employees must accept the industrial espionage prevention program and cooperate to make it work. If employees do not believe in it and do not cooperate in it, the program, like any other security control, is undoubtedly doomed to failure. The key to acceptance is understanding and knowledge of the program and the ways in which valuable data can be lost. Education must, therefore, make employees aware of the need for the program and motivate them to become part of it and help make it work effectively. If the education is effective, no employee should be able to say he does not understand the operation of the program or that he does not understand the responsibility assigned him to make the program work.

Communication methods must be established to insure that employees are being properly instructed. The education process may be formal or informal depending upon the size of the organization and the number of employees. Information pipelines that have already been established are usually the most effective ones because employees are already familiar with them. Many organizations have published booklets that outline the essential elements of the industrial espionage prevention plan. Company newspapers, posters, coffee cup coasters, lectures, and discussions in regularly scheduled staff meetings have also been used effectively.

In summary, regardless of how it is conducted, the education process should be designed to accomplish three things: (1) give each employee a full explanation of the reasons why the industrial espionage program is required, including the hazards with which the organization and the employee must cope, (2) provide a full explanation of the operation of the program, and (3) give each employee an explanation of the specific responsibilities he must assume for the protection of the organization's sensitive information.

Physical Controls

Physical security controls are an important aspect of the industrial espionage prevention program because they can limit the access of people to material and information that might be of value to a competitor.[3] But although physical controls are of great value, they cannot be expected to do more by themselves than discourage the undetermined and delay the determined. They alone cannot, that is, be relied on to prohibit access; they must be used in connection with the other prevention techniques described in this chapter. "Security in depth" is a term commonly used to indicate a series of controls. The purpose of the series is to set up enough obstacles to make an attempt to penetrate a facility or area too unprofitable or too risky.

An anonymous professional business spy who described his activities in a national magazine in the mid-1960s related how he had been able to steal oil exploration maps worth millions of dollars because the victim company had not followed the security-in-depth principle when it designed its facility protection plan. The spy was able to bribe a half dozen armed plant guards to let him in a side door, and he was then able to enter an unsecured area and photograph the maps. "To this day," he said "that company thinks that my client simply got to the oil first because of better researchers."[4]

In that instance, the victimized organization had relied entirely on guards to protect its sensitive material. Guards are certainly of value in most security plans, but the lesson to be learned is that they cannot be relied on for complete protection. Other security techniques must be related to their activities, and apparently the oil company management was too naive, complacent, or careless to adopt them. Because the company representatives did not become aware that they had been victimized, the security of the facility may still be in the same vulnerable condition.

Some examples of physical controls are the following: fences and other outside barriers; guards and receptionists; lights, locks, and electronic alarms; doors, turnstiles, gates, and window barriers; and vaults, safes, and other compartments of special construction. Physical security controls to limit the activities of people are important in the overall program because, as was pointed out in Chapters 4 and 5, trade secrets are lost through the activities of individuals both within and outside the organization. For that reason, the controls discussed in this section will be divided into two groups: external and internal.

EXTERNAL CONTROLS

The purpose of external controls is to prevent nonemployees from gaining access to the facility or to critical areas within the facility. Included are visitors, customers, and anyone who might want to gain entrance in order to cause harm. When external controls are considered, the facility should be regarded as a box with not only sides but also a top and a bottom. Barriers and such protective measures as alarms, lights, and guard patrols should protect the roof of the facility and also underground areas such as basements and sewers. As to the walls, protection must be provided for not only windows and doorways but also other openings that might be used to gain access.

Visitors, customers, and other nonemployees will from time to time need access to the facility, but their movement inside the facility should be controlled. The first step is to provide a receptionist or guard. Anyone who claims to have a need for access can then be screened at the entrance to determine if he really does have a need. If he does, he should be required to register on a visitor's log. The log should be so designed that the visitor gives his full name, identifies the organization he represents, and records his times of entrance and departure. Many organizations also require the individual to identify himself by means of a company ID card with his photograph or some other relatively positive means. If the enterprise is engaged in government work, the visitor is usually required to state his citizenship on the log.

Having been properly identified and checked into the facility, the individual should be escorted while he is there. Employees who escort visitors should be instructed to allow them to have access only to areas pertinent to their visit.

The visitor's log should be in the form of individual cards or sheets, one for each visitor. Some organizations use a daily sheet but that is regarded as undesirable because a visitor to the facility late in the day has the benefit of seeing the names and affiliations of all preceding visitors. The information might be of value to him. For that reason, most orgnizations that operate in a competitive posture have adopted the single card register that is removed from the reception desk and filed by the receptionist as soon as the individual completes it.

If an outsider is to visit an area in which he might obtain sensitive data, he can be asked to sign an agreement that he will not disclose or use any information he may obtain as a result of his visit. Such an agreement might be in the form of a letter addressed to the organization being visited and signed by the visitor:

> I agree to hold in confidence all information given to me during my visit to the _____ laboratory of _[company]_ on _____ and in particular information given to me regarding the development of_____ . I further agree that I will not divulge or utilize such information improperly. This agreement shall not apply to any information that becomes publicly available.

Some companies have incorporated such an agreement into the visitor's log. When the visitor signs the log, he certifies that he has read the statement concerning sensitive information and also certifies that he will not divulge or misuse any information he may obtain as a result of his visit.

INTERNAL CONTROLS

Internal controls are intended to limit the access of nonemployees who have entered as visitors and also trespassers who have gained access without permission. In addition, they protect against the curious or dishonest individual working inside the facility who might wish to remove sensitive material or information and against any undesirable activities of consultants, outside contractors, and others who need free access to the facility in the way employees do. All outsiders who are regularly within the facility should be required to become familiar with and abide by all security controls.

Safes, vaults, and rooms or areas segregated from the rest of the facility can be utilized as physical controls to protect sensitive information.[5] Compartmentation of information and areas has been used for many years by the intelligence community, and internal physical controls can be used effectively to accomplish such compartmentation.

The simplest type of internal control is a locked container or cabinet. A large variety of such containers are available on the open market. They range from cabinets with simple key locks, which are not recommended for the protection of valuable information, to safe-type cabinets of heavy construction that have combination locks and are designed to give protection against forced entry, fire, smoke, and moisture. Still other cabinets are designed especially for the protection of classified government material. The value and importance of the material to be protected should determine the type of container selected.

A locked container is often called a safe, and so it is commonly assumed that anything stored in it will be properly protected. Unfortunately, most containers that are called safes give protection against either fire or forced entry, but not against both. Commercial safes are divided into two general categories: those designed to protect records and those designed to protect money or negotiable instruments. Because a container of the first type is constructed of light steel and insulation primarily to give fire protection, it will offer very little resistance to forced entry. A container of the second type, designed for protection against burglary, will not protect against fire because its thick solid steel walls will rapidly transfer heat to the interior.

It is possible, however, to obtain a security cabinet that has been designed to meet federal specifications for safeguarding classified material. It offers fire protection as well as security against robbery and burglary, but it is not usually available to commercial organizations that are not involved in safeguarding government classified material.

It should be noted that a common key-locked cabinet will not give effective protection because the lock can be picked or unlocked with any one of the many duplicate keys available for each cabinet manufactured. If that fails, the cabinet can easily be forced open, and so it cannot be expected to act as more than a psychological deterrent.

Vault-type rooms are also used, but they are commonly designed to protect only against fire. If so, protection against forced entry should be incorporated. For example, an alarm system might be included to signal a penetration, or a heavy container, such as a money safe, might be included to protect highly sensitive company information.

Rooms or areas are often segregated from the rest of the facility as a means of internal control of personnel. Arrangements are made to limit the number of individuals, other than people who work there, who have access to the area, usually on a need-to-know basis. The area may be a small room in which a very few people work, or it could even be an entire controlled building in which sensitive data are handled. Segregation control is effective in providing physical protection for reproduction, art and design, and prototype and model areas; laboratories in which sensitive work is being done; planning centers in which charts and other sensitive data are constantly exposed; or any other areas to which access should be restricted. Area control is relatively inexpensive and will generally give effective security protection. There is no limit to the number of controlled areas that can be planned into a facility. Further safeguards or inner rings of protection can be planned into the areas. For example, alarms can be installed for protection when areas are unoccupied, or locked containers can be supplied within the area.

The control of the entrance to a segregated area is important. A number of doors may be required by fire and safety regulations, but only one of them should be an entrance. The limitation is necessary so that those entering and leaving can be controlled. Doors installed to meet fire and safety requirements can be equipped with alarms and designed without hardware on the outside, which will secure them from that direction. Panic hardware and locks can be designed into the doors on the inside to provide for opening in an emergency.

Two periods should be considered when controlled areas are established: one when the area is being used and the other when it is not. When an area is being used, the entrance can be secured with a lock and controlled by an individual in one of a variety of ways. If the area is a relatively small one, an individual on the inside might be responsible for controlling personnel coming and going. If the area is too large to be controlled in that way or if the traffic at the entrance is too heavy for such an arrangement, a clerk or a typist working on nonsensitive material might be placed outside the entrance and be assigned the additional duty of controlling the entrance. An electrically operated lock on the door might be activated by the typist, who would also monitor entrances. Alternative controls can be as elaborate as cost will allow and will no doubt also be determined by the importance of the work being conducted inside. Regardless of how the area is controlled or by whom, an access list of all those authorized to enter is essential. It should be in the possession of the individual charged with controlling the entrance and be kept up to date and accurate.

In addition to the access list, control techniques may include the use of badges that have been coded to indicate the area or areas to which individuals may have access. Some organizations have used colored jackets or smocks to assist in the control of individuals.

Information Control

The control of information is an important aspect of any industrial espionage prevention program. Items that should be included in the control process are blueprints, computer tapes or files, recordings, models, formulas, and papers or documents. The control of data that are transmitted orally, such as telephone conversations, discussions in meetings, and speeches, should not be overlooked. The Appendix provides an example of a model control practice that outlines general requirements for the distribution and control of information. It may be useful as a guide in developing a practice to meet the requirements of a particular organization.

Information that is to be protected must first be defined; otherwise, a large amount of material that is not really sensitive will be controlled. If that is so, the efficient operation of the system will be hampered and money will be wasted in protecting material unnecessarily. Also, employees will be less likely to accept the program and cooperate fully if they find that material that does not require protection is included.

After the information to be secured has been defined, it must be marked so that all recipients will be aware that it is to be protected. (The principle of classification and the need to adopt an appropriate designation are discussed in Chapter 7.) Usually, each document or other medium that contains classified information is rubber-stamped with the classification. Stamp letters should be large enough to be distinctive, and the stamp imprint should be in a conspicuous place on the sensitive material. Some organizations add a warning legend such as the following one:

This document contains information of a proprietary nature. All information contained herein must be kept in confidence and is not to be divulged to other than employees of the company who have a need to know and are authorized by the nature of their duties to receive such information.

Cover sheets for documents are also used to designate sensitivity and to instruct readers in proper document protection. See the Appendix for an example of such a cover sheet.

The release of information is a most important aspect of information control. The basis for release is, of course, the need to know, and the responsibility for determining who has a need to know must be assumed by the individual who generates or transmits the information.

If information is to be transferred outside the organization, additional controls should be available to insure that the data are properly protected. If sensitive information is to be removed from the company premises, the specific permission of a top management representative should be required. For example, a number of companies have adopted the rule that sensitive company information is never to be taken to the home of an employee. Management control on material leaving the company is an essential protection element; because when the material is not on the premises of the organization, the risk of loss increases significantly.

All material to be published, whether in a trade journal, sales brochure, or press release, should be carefully reviewed and approved by appropriate top management representatives to insure that no sensitive data are included. Any information to be released by an individual, such as a paper to be presented at a seminar or an article submitted for publication, should be reviewed and approved before it is released. Displays or exhibits to be sent out of the company should be carefully checked to insure that sensitive data are not included. The Appendix shows an example of a policy statement that deals with the release of information.

Individual employees who are in possession of material that has been marked as company sensitive should not be allowed to reproduce it without specific authorization. Many organizations require that supervisory approval must also be given for each reproduction.

The proper safeguarding of sensitive information during transmission must also be considered. Within the organization, the transmission may be in a sealed envelope that is marked with the classification of the material enclosed. Double envelopes may also be used. The envelope that contains the sensitive information is marked with the classification of the information and is placed in a second envelope that is sealed and addressed as if it were to be placed in the U.S. mail. There is no indication on the outer envelope that sensitive information is being transmitted.

A document control system can also be set up; its purpose is to insure that all copies of a document are accounted for. The system consists of a numbering arrangement and a log in which the title, number, and location of each document are recorded. A periodic audit must be conducted to insure that each copy is in the hands of the individual to whom it was sent and to insure that none of the copies have gone astray.

Regardless of whose hands it may be in, the material should be placed in a locked container when it is not in use. The selection of the container was discussed earlier in this chapter under Physical Controls. A container appropriate for safeguarding the material may be assigned to the user, or the user may be required to turn in the material each day so that it can be secured in a central file or vault.

When information is to be transferred outside the organization, as to a subcontractor, vendor, or some other organization, the outsider must be informed that the material is sensitive. In Chapter 7 it was mentioned that an agreement might be obtained from the outside organization to certify awareness of the sensitivity of the information and promise to give it proper protection. Needed in addition to an agreement is a determination that controls are adequate to protect the data while in the possession of the outside organization. An agreement to protect without a corresponding ability to safeguard would, of course, be worthless.

A secure way to destroy material that has served its purpose is an important element in the control of information. As was pointed out in an earlier chapter, spies consider trash an excellent source of information. For that reason any document, model, blueprint, or like material should be so disposed of that an industrial spy cannot obtain it and utilize it to the detriment of the organization. Sensitive data should never be disposed of in wastebaskets or in the trash, and the method of collecting such material should be incorporated into the industrial espionage prevention plan. Burning is an effective way to destroy most material. Many organizations utilize shredders or pulpers to dispose of paper that contains sensitive data.

Information that is transferred verbally must also be controlled. A rule that sensitive information is never discussed on the telephone should be adopted. Also, before any data are discussed with outsiders, in meetings or otherwise, the necessary management approval should be obtained and those who receive the information should be required to sign an agreement of the type discussed in Chapter 10.

Codes have often been used to protect information from competitors, consumers, and employees.[6] The technique is not new, because the Knights Templars of the Middle Ages enciphered letters of credit that the members carried instead of cash between the 9,000 Templar commanderies in Europe. The Templar cipher alphabet was based on the Maltese cross. Today most organizations rely on restricted distribution and careful supervision of the codes to protect them. Highly competitive businesses such as oil, mining, and banking often use private codes. Commercial cipher machines that provide a high level of protection of information encoded on them are available from several manufacturers, but most organizations depend on simple codes and ordinary security precautions.

There are recorded instances of a code system being used against the organization that originated in. One example involved the misuse of a code by an employee of the Hollandsche Bank Unie in Haifa in 1958.[7] David Hermoni, one of three employees who knew the bank's private code, opened two accounts in a Zurich bank. He cabled three banks in New York in code and instructed them to transfer $229,988 to the two accounts. He then flew to Zurich and, after identifying himself, withdrew $150,000 from one account and $50,000 from the other. Later he returned to withdraw an additional $25,000. In the meantime a confirmation cable from one of the New York banks to the Hollandsche Bank Unie in Haifa had alerted Hermoni's superiors to the fraud. Instead of getting the last $25,000, Hermoni was arrested.

Another way to protect information is to create a missing information link by limiting the knowledge of one element in a formula or process to a very few top officials of the organization. Probably the best known application of the technique is that by The Coca-Cola Company. The success of that business giant is based on a secret ingredient developed in 1886 and known only to two employees. They periodically go to a highly protected laboratory and mix batches of the key ingredient, which is known as 7-X.[8]

Personnel Policies

Preceding discussions of the release of sensitive information have indicated that the morale of employees is an important factor to consider when the industrial espionage prevention program is designed. Proper compensation of the employee who handles sensitive data is an important aspect of insuring that employee's loyalty to the organization. It must include not only financial remuneration in the form of salary, bonus, and fringe benefits but also appropriate additional recognition. For example, a number of companies have adopted the policy of making cash awards to employees who have developed new and useful ideas, techniques, or trade secrets. The awards may amount to a few dollars or many thousands, and they may consist of a one-time cash payment or a royalty arrangement. Some companies recognize such efforts with awards other than money. Whether monetary or otherwise, the award is usually regarded as a token and is not intended to be related to the actual value of the idea. That is, the award is not intended to be payment in full for the information developed; its purpose is to encourage employees to bring ideas to the attention of the organization and to protect their own ideas and company secrets from competitors.

Other techniques of recognition may also be used instead of or in addition to monetary awards. One company presents a plaque that can be hung in the employee's office or home and also adds the employee's name to a permanent company honor roll that is displayed conspicuously in company headquarters. Another company has established an honorary club named after a former president. To be elected to it, an employee must have made a significant contribution to the scientific or technical growth of the company. Luncheon or dinner meetings sponsored by the company are held periodically. Those who attend receive publicity and special recognition.

Most organizations adopt the attitude that the personal life of an employee is his own business, but the outside business activities of employees who are involved in handling company-sensitive data are important. Any activity that would represent a conflict of interest should be discouraged, and the guidelines adopted should specify the limitation on the employee's outside activities. For example, the moonlighting of employees should be covered in the guidelines. Will moonlighting be allowed and, if so, under what conditions?

Some companies have adopted questionnaires as a means of checking on the outside activities of employees involved in the handling of critical data. The employee is required to complete the questionnaire periodically and certify to its accuracy. Most such questionnaires of this type require the employee to list any of his outside activities that might be in conflict with either his work in the organization or the activities of the enterprise. Once completed, perhaps annually, the questionnaire should be carefully reviewed by a top management representative so that any activities that might be of concern can be investigated further.

Personnel policy should provide for correction when procedures have been violated. The guidelines that are adopted should provide for all disciplinary actions to be taken within the supervisory chain. A violation should be referred to the appropriate supervisor so that the problem can be handled in the same way that any other employee performance problem is handled. Employees should be made aware that disciplinary actions can result from infractions of the established rules.

The terminating employee must also be considered in the planning of an industrial espionage prevention program. He should be asked to sign a statement in which he certifies that he is not taking any sensitive information with him and that he is aware that he is neither to discuss such information nor utilize it in his new employment. An exit interview is also often conducted with the terminating employee, and at that time the same points are stressed. The following is an example of a statement that terminating employees might be required to sign.

I certify that I have surrendered all company private data in my custody; that I will not knowingly and willfully communicate, deliver or transmit, in any manner, company private information to an unauthorized person, company or agency, that I will report to ___[company]___ , without delay, any incident which I believe to constitute an attempt to solicit information by an unauthorized person; and that I (have) (have not) (strike out inappropriate word or words) also received a termination oral security briefing.

In addition, a letter may be written to the new employer pointing out that the departing employee has been involved in handling sensitive company data. The letter may also define the general areas in which the individual has been involved. Such a letter is valuable because it not only puts the new company on notice that the former employee has had access to company information but alerts the management representatives of the hiring organization to the actions necessary to insure that sensitive data are not solicited from the new employee.

Employees who terminate because of retirement should also be considered in the development of personnel policies. It is not unusual for an organization to retain as consultants key retired employees who have been engaged in sensitive work. An agreement and a monetary arrangement can be designed to keep the retired employee from making his services available to competitors.

REFERENCES

1. For a more complete discussion of this subject, see R. J. Healy and T. J. Walsh, *Industrial Security Management: A Cost-Effective Approach* (AMA, 1971).
2. *Time*, February 21, 1972, cover story.
3. For a more complete discussion of the application of physical controls, see Richard J. Healy, *Design for Security* (New York: John Wiley & Sons, Inc., 1968).
4. "How I Steal Company Secrets," *Business Management*, October 1965, p. 3.
5. For additional information concerning safes and vaults, see the following: *Standards for Protection of Records*, No. 232, National Fire Protection Association, Boston. *Burglary Resistant Safes*, UL 687; *Security File Containers*, UL 505; and *Fire Resistance Classification of Record Protection Equipment*, UL 72; all three published by the National Board of Fire Underwriters, Chicago. *Burglary Insurance Manual*, National Bureau of Casualty Underwriters, New York.
6. David Kahn, *The Codebreakers* (New York: The Macmillan Company, 1967).
7. Ibid.
8. William Bowen, "Who Owns What's in Your Head," *Fortune*, July 1964, p. 177

10

---◆---

AGREEMENTS
FOR PROTECTION OF
SENSITIVE INFORMATION

Industrial espionage cannot be prevented by the use of agreements, but the vulnerability can be considerably lessened. The determined spy or the hostile employee will steal information if given access to it, and all of the precautions in the preceding chapters are directed at limiting such access to the fewest persons consistent with the needs of the enterprise. The persons to whom access must be granted are those for whom protective agreements are intended. By establishing duties and accountability among persons with legitimate access, the likelihood that such persons could successfully make unauthorized use of the sensitive information is reduced. Also reduced is the chance that negligence by such persons in the handling of sensitive data would go undetected. By upgrading the conscious sense of responsibility among those granted access, the whole program of protection is made more effective.

Three among the agreements that are used to protect sensitive information are most common. The first is the simple secrecy or nondisclosure agreement. Next is the noncompetitive or negative competitive covenant, an agreement not to work for a competitor or engage in competitive activities. The third is the patent agreement and assignment of inventions. The last is used very widely in the United States even among businesses that do not depend heavily upon trade secrets. However, it is directly related to the protection of proprietary information and will be considered in this chapter.

Secrecy or Nondisclosure Agreements

Its proprietor will reveal sensitive information to two general groups: employees and nonemployees. Each group has a legal status that is quite different from that of the other, and each requires different treatment if disclosures to its members are to be protected.

EMPLOYEES

Whether or not an employee executes an agreement to protect his employer's sensitive information, he is under a legal duty to do so because of the special relationship of trust and confidence—a fiduciary relation in the eyes of the law—generated by the employment. That duty requires him to do or refrain from doing the things that, in the circumstances, a reasonable man would do or refrain from doing to avoid an unauthorized disclosure of the employer's confidential information.

At first reading, that seems to give an employer all the protection he needs. But if the employer relies only on the legal duty of his employees, all the elements of a breach will be questions of fact to be determined in a judicial proceeding. Was what the employee did or did not do a contributory cause to the loss of information? Would a reasonable man in the circumstances have acted the way the employee did? Most important, was the information disclosed a trade secret? Note carefully that not all kinds of sensitive information will be protected by the legal duty imposed on employees; only the kind that meets the legal test for trade secret is covered. An employer who relies on his employees' duty not to disclose trade secrets must not only prove the facts of the disclosure but also litigate the status of his claimed trade secrets. Failure to prevail—and the law is not clear in many areas—could result in a substantial loss to the employer.

Moreover, no single fact or circumstance will determine the trial of the issues. All the facts will be examined by the court or jury in an effort to reach an independent decision that a legal duty was culpably breached. An act or omission that might be clear evidence to the employer that the employee breached his duty might not convince the court. Because of those difficulties there is little wisdom in an employer relying exclusively upon his employees' legal duty to protect sensitive information disclosed to them.

A formal, express agreement is extremely valuable for the following reasons: (1) It is clear evidence of the employee's awareness that sensitive information exists and may be disclosed to him. (2) It can specify particular acts to be performed or avoided by the employee. (3) It can establish a basis for identifying the disclosures that are sensitive. (4) It can be the means for proving that an employee was aware of the sensitivity of even inadvertent disclosures made to him. (5) It can be extended to information that does not qualify as a trade secret and would not be protected under the general fiduciary duty. (6) It can establish a different legal framework—for example, a contract instead of a tort action for breach with different statutes of limitations and measures of damages, or it can substitute arbitration for trial.

Violation of an express agreement by an employee not to disclose specified kinds of proprietary or sensitive information would require no more than proof of the agreement and the disclosure to the employee together with the unauthorized actual or threatened further disclosure by the employee. General practice, what "reasonable men" might do or not do, and the inherent nature or value of the information would be irrelevant. The employee would be bound by his express agreement. His threatened disclosure could be prevented, and his actual disclosure would be legally remediable.

Even here, however, there is need for caution. Although an actual disclosure by an employee in violation of his agreement would give the employer a cause of action for damages, that might not be adequate for the employer in a practical sense. If the disclosure resulted in third persons innocently gaining knowledge of the information and if the employee had no personal assets adequate to compensate the employer for the damage caused by loss of the information, the action might more nearly resemble a penalty against the employee than redress of the wrong against the employer. Nevertheless, the availability of such action will be a deterrent to the employee and will afford the employer greater protection than he would have without the agreement.

The Appendix offers model provisions that should be considered for inclusion in an employee secrecy or nondisclosure agreement. Such an agreement can stand alone, or it can be included in a broader document that might also cover negative competitive covenants and invention assignments, both of which are discussed later.

NONEMPLOYEES

Unless there is some special relationship of trust between the nonemployee party to whom sensitive information is disclosed and the enterprise making the disclosure, there is no duty to protect the information in the absence of an express or implied agreement. For example, if (1) a business firm takes a drawing or a set of plans or a detailed specification to a vendor or supplier for the purpose of obtaining some part or product from the supplier, (2) the plans contain and disclose trade secrets or other confidential information, and (3) there is no agreement by the supplier to safeguard it, that information has been compromised and is thereafter public domain.

The agreement to safeguard could be implied by the conduct of the supplier and by all the facts. Thus, a formal-notice letter to the supplier of the nature of the information in the plans, plus appropriate markings on the plans or labels on models or nondocumentary materials, plus performance or partial performance by the supplier without any notice of his exception to the restrictive nature of the disclosure to him would probably establish an implied agreement to safeguard the information. However, the specific way in which safeguarding would be accomplished would be determined after the fact by application of the usual standard: what would a reasonable man have done in the circumstances?

In given cases, extraordinary precautions may be necessary for adequate protection of the sensitive information entrusted to the outside vendor or supplier. If reasonable precautions were taken but the information was nevertheless compromised owing to particularly persistent efforts of third persons to obtain it, there is a serious question whether any agreement binding the vendor to apply extraordinary precautions would be implied.

The use of a formal agreement with outsiders is therefore highly desirable for several reasons. First, it establishes beforehand, or at least contemporaneously with disclosure, that the outsider is accepting the disclosure as a restricted one. That avoids the problem of possible unrestricted disclosure and introduction into the public domain. Second, it permits the recitation of specific protective steps and routines to be followed by the outsider and does not leave the matter to a later determination of what would have been reasonable under the circumstances. Third, it permits specific exceptions of certain kinds or classes of information from the agreement. Normally such exceptions are demanded by suppliers and other commercially active disclosees for information that (1) the proprietor himself later makes public, (2) the disclosee had previous knowledge of, or (3) the disclosee gains later knowledge of through legitimate, unrestricted sources.

In addition to the express agreement with the outsider, supportive procedures should be followed to make enforcement of the agreement possible. The following are among the more important of such procedures and should be provided for in the agreement itself:

1. The information furnished the outsider will be appropriately marked or labeled at the time of disclosure. Reliance upon oral disclosure accompanied by oral warning presents later problems of proof. Although the creation and release of documentary information may increase the risk of exposure of the data contained, that risk seems preferable to the risk that mere oral disclosure or warning might not later be provable in a contest. The documentary release risk can, of course, be further restricted or reduced by proper use of the rest of the precautions in this list.

2. Each release will be documented and acknowledged by the recipient. A signed receipt is the best evidence.

3. The requirements for protection in use, storage, and transmission will be specified, as, for example, by incorporating by reference the proprietor's own security procedures, receipt of a copy of which the outside vendor acknowledges, or by attachment to the agreement of a specific schedule of the precautions to be followed by the vendor.

4. The outside vendor will be required to warn those of his employees to whom disclosure must be made in furtherance of the stated purpose of the agreement of his obligation to the proprietor of the information and his employees' duty to him in that regard. A record acknowledgement of such briefing is desirable from each vendor employee to whom disclosure is made. (How far an out-

side vendor will go in agreeing to these administrative requirements will, of course, depend upon the bargaining position of the parties. A proprietor of trade secret and other confidential information who is serious about protecting it will seek as much protection as he can get.)

5. The vendor will return to the proprietor, either immediately after it has served its purpose or upon termination of the agreement, all sensitive information released to him and all copies he necessarily made of it while it was in his possession. The return should be accompanied by a closing statement, as of the date of the termination of the agreement, that the vendor has returned or destroyed, as provided by the agreement, all sensitive information released to or reproduced or developed by him in the course of the agreement. The renewed certification is desirable even though the same representations are contained in the original agreement.

There are a considerable number of outsiders to whom disclosure of sensitive data may be required, and the exact terms of the agreement will vary with the nature of the other party. Disclosure may be made to vendors of supplies or services, consultants, licensees under patent and other arrangements, potential joint venture associates, candidates to acquire or be acquired in a merger, and even, in some circumstances, candidates for employment in highly specialized categories. The basic objective of the agreement, irrespective of the category of the person with whom it is made, is to secure the prevention of unrestricted disclosure and to achieve, to the maximum possible extent, effective protection of the information in the hands of the outside party. The sample nonemployee secrecy or nondisclosure agreement in the Appendix sets forth useful language that could be included, modified as required, in such a document. There are a number of other extremely informative sources of model agreements that also should be consulted.[1]

Noncompetitive Covenants

One of the surest ways to avoid the use by or disclosure to a competitor of trade secret and other sensitive information is to prevent an employee from taking employment with the competitor after he acquires the information. The most effective way to prevent that employment is to have express agreement between the employee and the original employer. The agreement, variously referred to as a negative or noncompetitive agreement or covenant, may stand alone or be incorporated with other terms and conditions as part of a broad agreement that covers all or most of the entire employment relationship. It is common, for example, to find a single agreement combining the nondisclosure covenant previously discussed with the noncompetitive covenant, which is under consideration here, and the patent and invention assignment, which will be treated later. Moreover, it is not necessary—although it is certainly possible—to include terms generally regarded as constituting an employment contract.

Thus an employer can have an employment "at will" or "at sufferance," meaning terminable at any time the employer chooses or unilaterally variable as to terms and conditions such as wage or salary or working hours, and still have an enforceable agreement as to noncompetitive future employment. That can be achieved by providing a separate consideration for the noncompetitive agreement, such as scheduled or determinable payments by the former employer in circumstances that prevent the employee from taking new employment because of the agreement. The model agreement in the Appendix contains such an additional consideration. However, it is possible in some jurisdictions to obtain enforcement of a noncompetitive covenant supported only by an employment at will.[2] In other jurisdictions the reverse will be true.[3] The safer course is to provide some additional consideration beyond a mere employment at will.

Because noncompetitive covenants are partially in restraint of trade to the extent that the potential new employer is denied the services of the former employee, there is a requirement that they be reasonable. It is well settled in the United States that a permanent agreement not to engage in competitive activity will not be enforced. Neither will agreements that are too broad as to the territory, activities, or kinds of competition they attempt to restrict. Moreover, the existence of actual trade secrets rather than confidential information other than trade secrets will be insisted upon in most cases in which extensive prohibition is sought.

Some states have declared it a matter of public policy that no person can be restrained in the practice of a profession or calling, and in such states the noncompetitive covenant is generally unenforceable.[4] However, even in such states, given a clear showing of trade secrets and a danger that they will or may be divulged, some injunctive protection may still be obtained or the agreement may still be enforced in a limited way.[5]

The objectives of a noncompetitive agreement should be these:

1. To identify the kinds of employment activity that are to be avoided. Not all employment activity, even with a competitor, will seriously jeopardize the former employer's trade secrets or confidential information.

2. To identify the kinds of competitor with whom future employment is to be restricted. That can be by clear description of the industry, the function within the industry, or the nature of product or services that characterizes firms with which employment is to be restricted.

3. To identify the geographical or territorial limits within which the restrictions apply.

4. To establish the period or duration of time over which the restrictions apply.

5. To keep (1) to (4) as narrow or brief as is consistent with the real protective needs of the original employer. If a certain kind of technical information or trade secret is realistically likely to be legitimately available fairly widely (as by independent discovery or unrestricted disclosure from other sources) within two years following the termination of employment of an employee who has the knowledge, it would be unnecessary to seek a restrictive agreement for four or more years. With increasing mobility on the part of employees, courts will probably demand the least restrictive restraints and will strike down or modify agreements that seek more.

6. To provide some adequate consideration to compensate the restricted employee for the rewards, material and psychic, that he may have to forgo because of the agreement. Increasingly the employer who seeks comprehensive protection is being expected to provide significant consideration—more than the bare minimum or "legal consideration." If the protection sought is of real value to him the employer should be able to put a price of some kind upon it. More than token compensation required to enforce a noncompetitive covenant is therapeutic for several reasons. First, it will require a case-by-case determination of whether or not to seek enforcement. Second, it will tend to hold the period of restriction to that which is least expensive for the employer. Third, it will tend to sharpen application of basic defensive programs such as need to know very early in the employment relationship.

Although the serious employer will seek to be equitable in his demands and consideration, he will also seek to provide the restricted employee with an incentive to do more than make a claim for contract benefits. Thus a restricted employee should be required to make positive efforts to find suitable nonrestricted employment and to demonstrate his good-faith efforts as a condition to collecting under the agreement. The model agreement for former employees (see Appendix) includes a good-faith-effort provision.

It is apparent that no agreement will deter a former employee or a vendor or a person who supplies some outside service from compromising a company's trade secret or other confidential information if that is the positive intention. At best, agreements provide a basis upon which to seek redress or remedy. When there is access to sensitive information and a motive for using or disclosing it without authorization, any proprietor of data is vulnerable to loss. As in patent infringement situations, the victim may not even be aware that his data have been taken or are being used.

It should be crystal clear, then, that the best agreements are not a substitute for limitation of disclosure, careful selection of persons who will receive disclosures, and vigorous application of the other protective techniques discussed elsewhere in this book. However, when disclosure is really necessary in the best interests of the owner of the data, properly drawn agreements can reduce or eliminate the likelihood that persons who receive the disclosure will use the information improperly. The mere fact that such persons are accountable for disclosed information and are bound to observe the protective routines in regard to the information that, it can be proved, was disclosed to them in light of the agreement will lessen the gain that they might otherwise expect to derive.

Patent and Invention Assignment Agreements

Almost any employee in good faith will recognize the right of the employer to protect sensitive information that was developed without his help. He is apt to have different and much stronger feelings, however, when he does help develop information that the employer then claims as his own. In fact, unless the employer takes care to spell out a more extensive right, the law will usually prevent him from claiming title or exclusive use. A patentable invention or discovery made by an employee in the course of his employment would normally belong to the employee. The employer would have only shop rights, that is, the right to use the invention on a royalty-free basis. Title to the invention and the right to patent would be the employee's.

Ownership of the patent by the employee would entitle the employee to license others to use the discovery. Additional licenses could work a major competitive disadvantage to the employer and conceivably even put him out of business. To protect against those contingencies, many employers require all or a selected group of their employees to execute a patent agreement and assignment of inventions as a condition of employment. The agreement specifies that during the course of employment (and often for a specified term following it) any invention or discovery made by the employee with regard to any product or process or business of the employer, and irrespective of whether the invention was made on employer time or with employer materials, belongs to the employer.

The employee usually also agrees to do all things and execute all papers necessary to permit filing a patent application if the invention or discovery involves a patentable idea. Most agreements, however, are not restricted to patentable ideas and give the employer complete right and title to any idea developed during or following the term of employment that relates to the business of the employer.

The reason for extending the application of the agreement for a period of time after termination of employment is to avoid the situation in which the employee realizes that he is making or has made a valuable discovery, quits the employer, and then sells his idea to the highest bidder. Although an employee might still suppress his discovery, the inability to make immediate use of it elsewhere will operate in favor of his disclosing it to the employer as was intended by the agreement. The Appendix contains a representative agreement.

In the case of a nonpatentable discovery or one that could, alternatively, be protected as a trade secret, an employee could suppress news of his discovery, employ a confederate or deal directly with an unscrupulous competitor, and sell the idea and so defeat the original employer's rights. Like all other vulnerabilities to loss of sensitive information, the vulnerability to loss of ideas and inventions, despite assignments, can be only partly neutralized by the agreement.

Agreements Protecting Former Employers

An employer may find himself accused or attacked by another party who has been the victim of trade secret or sensitive data loss. The victim may allege that one of his former employees, now with the accused employer, has unauthorizedly taken and disclosed to the new employer trade secrets belonging to him. Of course, the burden of proving that charge is upon the alleged victim. However, as an evidence of good faith and as part of the awareness training provided new employees, an employer can include among the agreements required an undertaking by the new employee that he is not disclosing and will not disclose in the future to the new employer any trade secrets or confidential information belonging to other persons, including former employers.

REFERENCES

1. R. Milgrim, *Business Organizations—Trade Secrets*, Rev. Ed. (Albany, N.Y.: Matthew Bender & Co., Inc., 1972). "Employee Patent and Security Agreements" (New York: Conference Board, 1965). *Warren's Forms of Agreements* (Matthew Bender & Co., Inc., 1971).
2. *Wark* v. *Erwin Press Corp.*, 48 F 2d 152.
3. *SuperMaid Cook-Ware Corp.* v. *Hamil*, 50 F 2d 830.
4. States that prohibit negative competitive covenants include: Alabama (19 Ala. Code 22, 23), California (Cal. Bus. & Profess. Code, 16600, 16601), Florida (542.12 Fla. Statutes), Michigan (Mich. Com. Laws 445.761), Montana (13 Rev. Codes 807, 808), North Dakota (Cent. Code 9-08-06), and Oklahoma (15 Okl. Stat. 217, 218).
5. *Allis-Chalmers* v. *Continental Aviation*, 255 F Supp. 645; *Delta Finance Co.* v. *Graves*, 180 So. 2d 85.

11

---◆·◆---

ETHICS OF
INFORMATION COLLECTION

The primary focus of the first ten chapters of this book has been on methods that every competitive enterprise should adopt to protect information. This chapter is devoted to another important aspect of data handling that should be considered: the ethics of information collection. To begin with, the reader might be asked whether industrial espionage is being practiced in the collection of information by his organization. Anyone who conducts his affairs according to the Golden Rule will be startled by such a question and will undoubtedly indignantly react by answering negatively. He will probably naively assume that such a thing could not happen in his organization, that the employees of the enterprise would not stoop to such practices. However, experience has shown that the assumption that portions of an organization or individual employees are not engaging in questionable practices cannot simply be made.

Standards of ethics change constantly. Just as norms of sexual conduct and race relations have changed profoundly in recent years, so business practices in general have changed. Early business tycoons had little or no regard for either ethical or legal limitations. Andrew Carnegie is reported to have admitted to all kinds of shoddy business practices and rationalized them by saying that they benefited the public. On the other hand, William H. Vanderbilt said, "The public be damned, I am working for my stockholders."

Officers of companies no longer feel that they have power that transcends the law as did executives at the time of Vanderbilt and Carnegie. Most modern executives want to do more than is legally required of them and make every effort to conduct the affairs of their organizations in an ethical manner. Experience in general indicates that good, ethical practices are a business asset and a reputation for sharp or questionable dealings is not. But although there has been a striking general improvement in business and industrial practices over the years, the executive who adopts a self-satisfied, complacent attitude could find to his surprise that questionable practices are in effect in his own enterprise.

Two shocking situations that developed in 1959 illustrated that questionable practices can develop in entire industries. They were the price-fixing scandal in the electrical industry and the rigged television quiz shows. What alarmed and angered the public, even more than the revelation of the scandals, was that some of the top executives of the companies involved said they were ignorant of the questionable practices of their subordinates.

The defendants in the electrical price-fixing case pleaded guilty to the charge that they had conspired over a period of eight years to fix prices, rig bids, and divide markets for electrical equipment worth $1.75 billion a year. Involved were 45 top executives from 29 corporations, including the giants in the industry. Chief Judge J. Cullen Ganey of the U.S. District Court of Philadelphia, when passing sentence in early 1961 on those involved, took note of the fact that employees throughout the organizations were conforming with the conduct of their superiors. The judge also observed, "one would be most naive indeed to believe that these violations of the law . . . , involving so many millions of dollars, were facts unknown to those responsible for the conduct of the corporation." The president of one of the largest corporations involved is reported to have said, "I don't take the position that I can wash my hands of it. My viewpoint is that this is a management failure."

Because there was no apparent violation of law, the television quiz scandal did not result in any legal actions. However, the questionable practices utilized became known worldwide because of the popularity of the programs, on which a variety of contestants received large sums of money for giving carefully rehearsed answers to difficult questions. One of the most popular contestants, Charles Van Doren, testified before a congressional committee in late 1959 that his winning $129,000 on the NBC quiz show *Twenty-one* had been arranged.

The two incidents, although not connected with industrial espionage, suggest that careful consideration should be given to an examination of the methods used to gather information in every enterprise to insure they are not only legal but ethical. Self-regulation is always more desirable than action that is forced by government regulation, legislative enactment, legal action, or the indignation of competitors or the public. As was discussed in Chapter 2, the collection of information in any competitive enterprise is vital and the majority of executives will agree that it is essential as long as it is made legally and ethically. The question, then, is when a practice is illegal or unethical.

It is relatively simple to determine what would be considered illegal in the area of information gathering. The discussion in Chapter 3 concerning that aspect of the problem will be useful to anyone who needs legal guidance. On the other hand, the determination of ethical practices is much more difficult. Despite the opinion of Humpty Dumpty, who said, "When I use a word, it means just what I choose it to mean," the application of the word "ethical" is not simple. Webster defines ethical as "of or relating to moral action, motive, or character; conforming to professional standards of conduct."

ATTITUDE OF INDUSTRY

A number of books have been written about the general subject of business ethics, but very few data about ethical standards as they relate to information collection have been available. A study reported in the *Harvard Business Review* in 1959 gave some indication of how business executives feel about the collection of data.[1] Its findings were that the majority of those included in the survey would utilize ethical means to collect data.

The report, which is summarized in the Appendix, was based on 1,558 questionnaires completed by business and industrial representatives. Only a few of those who responded to the questionnaire indicated that any formal information collection system was being utilized in their companies. That did not mean, however, that they considered competitive information to be of no value or unimportant to the operations, because a majority of the companies simply relied on informal methods of data collection. Most of the respondents indicated that company salesmen were their best sources of information about the activities of competitors. The second most valuable source was given as published information, and the third was personal or professional contact with executives of other companies.

Since the *Review* survey dealt with executive opinion and not with the standards of conduct for the collection of information, the methods that salesmen and others in the various organizations actually used to collect data were not recorded. Apparently the executives themselves were motivated by high ethical standards, and they would no doubt want their personnel to be also, but in the absence of proper guidance some of the methods used at lower levels might be questionable. Possibly management simply assumed that employees needed no instruction in the proper conduct to follow in the collection of information. Unfortunately, personal ethical codes cannot be relied upon, both because standards vary from individual to individual and because personal codes fluctuate with changes in individual ideas, attitudes, practices, and pressures. It has been said that few people share exactly the same ethical outlook.

Senator Edward V. Long (Democrat of Missouri), whose Senate Judiciary Subcommittee on Administrative Practices and Procedures conducted extensive hearings on privacy invasion and industrial espionage, was quoted in 1966 as saying, "Most industrial espionage is not done with the knowledge of the people at the top. Someone down lower gets overzealous in his desire to succeed." [2]

An article in *Dun's* in late 1971 indicates that the Japanese also have recognized the importance of salesmen and field representatives; for they have established a special school to teach them American customs and methods. Although the account does not mention information collection as an objective of the course, one might read between the lines and conclude that it would certainly be a natural inclusion. [3]

THE CODE OF ETHICS

If an information collection plan is to be ethical, the essential element is a standard or code. It is not, however, possible to outline one that would be useful to every reader or organization. Each enterprise must develop standards based on its overall operation and needs. The guidelines established should not be like the old Prussian idea of a citizen's rights: "Everything that is not expressly allowed is forbidden." Also, to simply establish a standard of conduct is not enough. The top management of General Electric, one of the corporations involved in the 1959 price-fixing case, had prohibited price agreements with competitors. Evidently executives at lower levels disregarded the directive.

One author who discussed a general code of ethics for business use included information that would be valuable to anyone who wished to establish a standard for information collection. He said:

No code can be looked upon as the product of a crash program, requiring a spurt of sustained effort only to be relegated to a secondary position once the fine words are on paper. To maintain and elevate the ideals and standards—and practices—of our free enterprise system requires constant effort.

One such publication was issued by the American Management Association in 1958 and is entitled *Management Creeds and Philosophies.* And the first point emphasized here is this: "The process of formulating the creed is often more valuable than the finished product." In other words, great benefits flow from the simple fact that those writing a code must sit down and think about ethical problems.

One benefit is illustrated by the experience of a fairly small firm I know which has about one hundred key employees. When this company set about writing a code, top management began by meeting with all these employees to get a better picture of the specific ethical problems faced by those operating on a lower level. Management was amazed. It seemed that everyone in the company had been facing a wide variety of serious ethical dilemmas which they had handled case by case without any guidance from above. Worse yet, most cases had been resolved in favor of the course that would produce the greatest short-run profit. Management discovered that a number of shocking practices had been prevalent because of two employee attitudes: (1) The company had always placed greatest emphasis on rewarding those who made the biggest contribution to profits. (2) Because the firm never had evidenced any great concern for ethical considerations, it seemed to most employees that cutting corners in order to maximize profits was a condition of employment—that was how the business operated; if they didn't like it, they could quit.[4]

The author also made the point that it is not always necessary to reduce ethical standards to writing. Key employees of small firms can constantly be reminded in staff meetings of the standards of conduct expected in information collection. Larger organizations, however, must reduce the guidelines to writing, because verbal policy statements are easily misunderstood or ignored. Size is always a problem—General Motors, for example, employs more people than live in Boston.

RESPONSIBILITY OF MANAGEMENT AND EMPLOYEES

A code itself will not guarantee high ethical performance. Top management must comply with it and thereby set an example to others in the enterprise. If the ethical standards of top management are high, that will usually be felt throughout the organization. If, on the other hand, those at the top have low standards, there is every reason to expect those at lower levels to be corrupted. If questionable practices in the collection of competitive information exist within the enterprise, they are likely to have an adverse affect on the industrial espionage prevention program. If top management motivates or even allows employees to engage in questionable practices in their dealings with those outside the organization, it is possible that the employees will operate the same way within the organization. If employees are encouraged or allowed to utilize illegal or unethical methods to obtain information from competitors, then management should not be surprised if the same employees steal information from their employer.

Even when ethical standards for data collection do exist, an employee will sometimes find himself in an ambivalent position. There will be a tendency to resolve each doubt in favor of the most profitable course. Every employee must constantly relate to his supervisors, his subordinates or fellow employees, the public, the laws, the government, and himself. When he attempts to determine a position, the law may indicate one course of action, business practices another, and his conscience still another. Accepted practices within an industry and what one can get away with and still keep competitors and others from complaining are not standards upon which the application of a code can be based.

Supervisors at all levels must share responsibility with top officers of the enterprise for implementation of the information collection standards that are adopted. Further, they must be delegated responsibility for seeing to it that the guidelines established are understood by individual employees and are constantly being followed. The need for each individual to understand the information protection program is emphasized in Chapter 9. It is equally important that employees understand the standards adopted for information collection if that phase of the program is to be successful.

THE EROSION OF PRINCIPLES

If the consideration of the use of unfair and overreaching techniques to gather competitive information is regarded in isolation from other aspects of American life, it will be easy to miss a point that is really central. The point is that we are showing tendencies everywhere of abandoning restraints upon the selection of means when the end is profit. Manufacturers build goods that are designed to require rapid replacement but merchandise them as sturdy and durable. Advertising, abetted by the communications media, regularly cajoles us to use substances or engage in activities that independent evidence has shown to be positively harmful to health and well-being. The reason for such deception is profit—the profit of the manufacturer, the profit of the retailer, the profit of the advertising agency, and the profit of the TV network. In short, the sole controlling criterion of an activity in the business community is often profitability.

If questionable practices are followed merely because they yield profit, how can the questionable practices of industrial espionage be singled out for selective condemnation? Surely the inconsistency and lack of logic will be apparent to any serious observer. The organization concerned about preventing espionage by its own people will look carefully at the kinds of conduct it condones. The total moral climate within a business enterprise and within the business community will have a greater deterrent effect on espionage than specific programs designed to deter it.

And even without express approval, the tacit approval of a management that enjoys profitable operations will be clear if there is no inquiry into the methods that produce the profits. In other words, as has been recently stated by some giant enterprises, social value is as much a factor in business decisions as economic value. There is a duty to the community and the commonwealth not to earn monetary profit at disproportionate social costs. The erosion of principles on which fair dealing must be built is a social cost. To permit practices that accentuate that erosion is to violate an admitted duty. In the frame of a much broader moral inventory-taking, the attitude toward industrial espionage must be reviewed to be sure that what is condemned in words is not really praised in deed.

REFERENCES

1. Edward E. Furash, "Problems in Review—Industrial Espionage," *Harvard Business Review*, November–December 1959. Used by permission. For additional detail, see Appendix.
2. "Ferreting Out the Spies in Business," *Business Week*, April 2, 1966, p. 51.
3. " 'Steak, Yes—Martinis, No,' " *Dun's*, October 1971, p. 71. Copyright, 1971, Dun & Bradstreet Publications Corp.
4. Luther H. Hodges, *The Business Conscience* (Englewood Cliffs, N.J.: Prentice-Hall, Inc., 1963.)

APPENDIX

Model Policy Statement on Protection of Sensitive Data

SAFEGUARDING COMPANY INFORMATION

POLICY

Company information will be protected by measures commensurate with its nature. Safeguards will be established to prevent disclosure of information which would be harmful to the Company. No employee will disclose any Company information to persons outside the Company unless the information has already been made public or approved for publication. It is the personal responsibility of an employee desiring to make public disclosure to be certain the information has been approved for publication.

SCOPE

This policy establishes the regulations which govern the classification of Company information and specifies the protective measures required for information which carries a Company classification or which is to be disclosed outside the Company.

DEFINITIONS

PROPRIETARY. A classification assigned to any Company information whose unauthorized disclosure could result in harm to the Company or be detrimental to Company interests and which must be safeguarded in accordance with this Policy and supplemental instructions.

Transmission, internal. The communication, by means under Company control, of PROPRIETARY information to Company employees or Company organizational units.

Transmission, external. The communication of PROPRIETARY information to persons not employed by the Company, irrespective of the means of communication, or to Company employees by means not under Company control. U.S. mail, public telephone and telegraph, express and air express and TWX network are means of communication not under Company control.

CLASSIFICATION OF COMPANY INFORMATION

The Company has established the classification PROPRIETARY to assure proper protection for certain types of Company information. This classification applies only to Company information and is entirely separate and distinct from classifications assigned for the protection of classified defense or Government information.

The PROPRIETARY classification shall be assigned to any information meeting the terms of the definition. Such information may include without being limited to:

Research and development. Any information relating to Company-sponsored research and development projects.

Financial. Pricing of Company products, operating statements, return on investment, profit margins, bid or proposal data.

Technical. Product specifications, process specifications, test results, formulations, performance characteristics and limits.

Marketing. Schedules of delivery or distribution, customer lists, assessments of market and share of market, pending or planned negotiations, customer analyses or reports on products.

Organizational information. Data regarding the opening, closing, expanding, or modifying of Company facilities, mergers and acquisitions, transfers of responsibilities or transfers of personnel.

PROTECTION OF PERSONAL DATA

To provide protection and assure privacy of personal information about Company employees, the Company has also established the designator PERSONAL AND PRIVATE, which is to be applied to the following kinds of information about employees: Medical records, detailed personal history statements, appraisal and merit review evaluations, individual salaries not fixed by schedule or negotiation.

ASSIGNMENT OF CLASSIFICATION

PROPRIETARY classification shall be assigned originally by the cognizant Manager. This is a nondelegable function. Subordinate positions shall refer to their cognizant Manager any material which requires or may require classification as PROPRIETARY.

Reproduction, abstracting, incorporation into other information or other communication of information already classified PROPRIETARY does not require the approval of a Manager for assignment of the PROPRIETARY classification. Such classification shall be assigned to the later information by the person incorporating or otherwise communicating the already classified PROPRIETARY information.

PERSONAL AND PRIVATE designator shall be assigned automatically to the kinds of information previously described wherever such information occurs.

RESTRICTIONS ON DISCLOSURE OF PROPRIETARY
AND PERSONAL AND PRIVATE DATA

PROPRIETARY information and PERSONAL AND PRIVATE information shall be disclosed only to persons who require access, knowledge or possession for the performance of duties to the Company required by their positions as employees or contracts or other agreements with the Company.

Disclosure of PROPRIETARY information to persons not employed by the Company but who have executed contracts or agreements with the Company to safeguard Company information must be approved, in advance, by the Manager of the Division utilizing the services of such persons.

Disclosure of PROPRIETARY information to persons outside the Company who have not executed contracts or agreements to safeguard Company information must be approved, in advance, by the President or his designee.

PROTECTION OF PROPRIETARY AND PERSONAL AND
PRIVATE INFORMATION DOCUMENTS

Marking. Documents shall be marked conspicuously on each page, top and bottom, with the legend PROPRIETARY or PERSONAL AND PRIVATE, except that PERSONAL AND PRIVATE documents consisting of more than one page need be marked only on the first page.

Numbering. PROPRIETARY documents shall be distinctively numbered by the unit or department originating them. A permanent record shall be kept of each PROPRIETARY document generated or originated, the number of copies, and the distribution of each copy. PERSONAL AND PRIVATE documents require no special numbering or recording.

Transmission and accountability. PROPRIETARY documents shall be transmitted internally in opaque, sealed envelopes on the outside of which has been clearly indicated the name, unit, and location of both sender and addressee. At the option of the sender, a receipt may accompany each PROPRIETARY document, to be executed by the addressee and returned to the sender. All persons sending or receiving PROPRIETARY documents shall maintain a log showing the title, number, and copy number of each document, the date of the transaction, whether sent or received and if sent, to whom and, if accompanied by a receipt, whether the receipt was returned. External transmission shall be in two opaque, sealed envelopes, both of which have clearly indicated the name and complete mailing address of both sender and addressee. The inner envelope *only* shall also be conspicuously marked PROPRIETARY. Transmission shall be by registered or certified U.S. mail or by commercial express using continuous accountability service. Receipts shall be used in every external transmission. PERSONAL AND PRIVATE information shall be transmitted in opaque envelopes or other containers on which have been clearly indicated the name and complete address of both sender and addressee. External transmission shall be by U.S. first-class mail, U.S. insured parcel post, or insured commercial express or air express. No receipts between sender and addressee are required.

Storage. PROPRIETARY documents, when not in use, shall be stored in approved vaults or safes or in steel filing cabinets secured by a special locking bar and padlock. PERSONAL AND PRIVATE documents, when not in use, shall be stored in a filing cabinet, desk or other storage container with a reasonably secure locking device. Documents left unattended and not under the direct observation of an authorized person shall be deemed not in use.

Reproduction. PROPRIETARY documents shall not be reproduced by any method if so marked. Such documents shall carry the conspicuous marking NOT TO BE REPRODUCED directly beneath the classification PROPRIETARY on each page thereof. Other PROPRIETARY documents may be reproduced only when specifically required for a particular, authorized purpose and only in the number so required. Each such reproduction must be approved by the supervisor responsible for the function or activity requiring the reproduction. All reproduced copies shall be entered in the Accountability Log of the unit or function making the reproduction and shall be assigned distinctive copy numbers. The disposition of each copy shall be shown in the log with the signature of the supervisor authorizing the reproduction. PERSONAL AND PRIVATE documents may be reproduced only in the numbers specifically required for a particular Company purpose. No record is required of such reproduction.

Destruction. PROPRIETARY and PERSONAL AND PRIVATE documents shall be destroyed by shredding or tearing in such a fashion that reconstruction of the original document is not possible. In addition, PROPRIETARY documents shall be burned or incinerated. Documents which can be delivered directly into a facility which will completely destroy them by burning and which remain under observation of an authorized Company representative until so destroyed need not first be torn or shredded.

OTHER THAN DOCUMENTARY MEDIA

Area controls. When, because of the size or nature of PROPRIETARY information in other than documentary form, it is not possible to prevent access to it by unauthorized persons, controls shall be established including posting of the area as PROPRIETARY and control of admittance through locked doors. Admittance shall be on a need-to-know basis and records shall be kept of the admittance of all persons not regularly assigned to the posted area. Cognizant Managers are responsible for advising the [Security Department or other named function] when area controls are required.

Transmission and destruction. PROPRIETARY information other than documents shall be transmitted and destroyed in such manner as the [Security Department or other named function] may prescribe in each case. Cognizant Managers shall request such guidance when required.

DECLASSIFICATION OF PROPRIETARY INFORMATION

Automatic. In cases involving PROPRIETARY documents, if the original classifier knows at the time of classification that the happening of a future event or the occurrence of a future date will make it unnecessary to treat such documents as PROPRIETARY thereafter, he may so indicate by including, directly after the classification marking, the legend: UN-CLASSIFIED AFTER [certain date or contingent event]. *Automatic declassification may not be used on* PROPRIETARY *documents containing technical information.*

All other cases. In all other cases, PROPRIETARY information may be declassified by the original classifier or by higher management in the same functional line. Declassification is accomplished by obliterating the PROPRIETARY markings on documents and by discontinuing controls on other media. Persons authorizing declassification shall communicate that fact to all record holders of documentary PROPRIETARY information. Declassification shall be accomplished in every case when it is clear that protective measures are no longer required. Record holders of declassified PROPRIETARY information shall note the declassification and its date in their respective accountability logs.

SUPPLEMENTAL INSTRUCTIONS

Local Company activities and facilities are authorized to issue supplementary instructions, not inconsistent with this Policy, to the extent necessary to accomplish the purpose of this Policy. Copies of all such local instructions shall be submitted to the [Security Department or other named function].

SITUATIONS NOT COVERED

Any situation involving PROPRIETARY or PERSONAL AND PRIVATE information which cannot be resolved by the application of this Policy or local supplementary instructions shall be referred immediately to the [Security Department or other named function] for appropriate instructions.

AUTHORITY

This Policy is issued by the authority of the President of the Company.

Model Information Control Practice

DISTRIBUTION AND CONTROL OF COMPANY PRIVATE INFORMATION

Certain information generated, received, and circulated within and outside the Company is of such a private nature that controls are required to prevent improper disclosure.

Information subject to this Practice is designated Company Private and is handled in accordance with the following procedure:

1. *The employee who originates or is the first to receive private information on behalf of the Company is responsible for:*
   a. Determination of the Company Private designation.
   b. Marking the document as indicated in paragraph 2 of this Practice.
2. *Marking*
   a. *Company Private information* is marked with a stamp bearing the legend "Company Private."
   b. The applicable marking is placed at least once on the face of the document or on the front sheet of a multipage document. Permanently bound documents are marked on the front and back covers and on the first page.
   c. In addition to the above marking requirements, a Company Private cover sheet may be attached to the document as a supplementary safeguard.
   d. In the event that it is not appropriate to place markings on a document, the applicable cover sheet will be stapled or otherwise securely affixed to the front of the document. Cover sheets and stamps are available in Stationery Stores.
3. *Distribution*
   a. *All distribution and disclosure of Company Private information* requires the determination, by any distributee, that a legitimate need exists for access to the information by the person to whom distribution or disclosure is proposed.
   b. *External distribution or disclosure of Company Private information* is not permitted except with the prior approval of a Corporate Officer.
4. *Transmission.* Company Private information transmitted within the Company by Mail Services is placed in a sealed, opaque envelope or container which is addressed and marked "Company Private." Messengers must not make delivery to unattended offices. Such information being hand-carried between company facilities or transmitted through the United States mail is, in addition, placed in an addressed but unmarked external sealed envelope or container; registry is not required.
5. *Storage.* The minimum storage requirement for Company Private

is a locked desk or locked cabinet. Such information must not be left unattended.

6. *Destruction.* Destruction is accomplished by sending the information to Classified Document Control in Destroy Envelope (Form 1050 or 1050A).

7. *Reproduction of Company Private information* is authorized by any distributee. The applicable designation must be stamped on all reproduction work orders relating to such information. Waste copies are placed in the burn barrel by Reproduction Services personnel, or destroyed as indicated in paragraph 6 of this Practice.

8. *Accountability or Document Control records* are not normally required.

9. *The Security Department* monitors compliance with the requirements for transmission and storage of Company Private information. The improper handling of such information is reported by the Security Department to the appropriate member of management.

Document Cover Sheet

FROM	TO	BLDG.	ROOM

INSTRUCTIONS—CORPORATE PRACTICE 6-80-1

TRANSMITTAL

1. Transmittal of private information within the Company is accomplished by placing the material in a single, sealed opaque envelope and attaching a cover sheet to the outside wrapper.

2. Transmittal of private information outside the Company to another company or agency is permitted only after specific approval has been granted by a corporate officer.

3. Externally originated Proprietary Information, being handled as Company Private, cannot be transmitted outside the Company without the approval of the Patent Counsel.

SAFEGUARDING

Minimum storage for Company Private information is a locked desk or locked cabinet.

REPRODUCTION

Reproduction of Company Private is in accordance with Corporate Practice 6-80-1 (Company Private Information).

DESTRUCTION

Destruction of Company Private information is by placement in destroy envelopes (Form 1050) and forwarding to the Destruction Office, Classified Document Control, Security Department.

Model Policy Statement—Publication of Information

PUBLIC DISSEMINATION OF INFORMATION

The Office of Information is the central point within the Corporation for the coordination of approval of information which may be publicly disseminated. This includes all press releases, photographs, motion pictures, advertising, brochures, exhibits, speeches or other information materials pertaining to the Company or its operations. Included also are survey questionnaires or queries from outside survey groups which seek information on Company business or operations, or on associate and/or prime contractors and subcontractors.

1. *Requests by news representatives* of the press, radio and television for information or interviews with Company personnel are referred to the Office of Information for coordination.

2. *Requests for recruitment advertising* are forwarded to the Personnel Directorate which coordinates the material with the Office of Information prior to release.

3. *Requests received by employees* for information about the Company, or for Company information which an employee wants to release

for publication, are cleared with the Director of the Office of Information prior to the employee making any commitment.

The presentation or publication of professional papers is handled through the Office of Technical Relations.

4. *Survey questionnaires or queries received by employees* concerning any of the above-mentioned or related subjects are forwarded to the Office of Information.

The Director of the Office of Information determines if a response to the query is appropriate. He is responsible for providing the Head of the Security Department with data necessary to keep the Military Security Office informed of the extent and nature of the activities of survey groups.

The Office of Information is responsible for determining that all matters pertaining to the release of information for public dissemination are appropriate, timely and in accordance with Company Policy and, when necessary, for obtaining the concurrence of the customer.

Employee Secrecy or Nondisclosure Agreement

In consideration of my employment by   [Company]   , and as a condition of such employment, I, the undersigned, agree and undertake as follows:

1. that   [Company]   may disclose to me from time to time, or I may learn in the course of my employment, certain secret and confidential information, including Trade Secret information belonging to   [Company]   ;

2. that all such secret and confidential information, including but not limited to information classified   [PRIVATE, PROPRIETARY, . . .]   remains the property of   [Company]   and is disclosed to or learned by me for the sole purpose of performing the duties of my employment;

3. that I shall not disclose any secret or confidential information belonging to   [Company]   to any person not employed by the Company without the prior, written approval of the Company, and shall not disclose such information to any other employee of the Company unless I know or have been informed by responsible persons that such other employee has a need to know such information for the performance of his duties;

4. that I shall protect and safeguard all secret and confidential information disclosed to or learned by me in accordance with the security

rules of   [Company]   as the same may be changed or modified from time to time, and that I have been informed of the rules in effect as of the date of this agreement;

5. that upon the request of   [Company]   , and in any event upon termination of my employment, I shall return to the Company all secret or confidential information in documentary or other tangible form, including copies or reproductions, then in my possession or under my control; and

6. that the obligation not to disclose secret or confidential information continues after the termination of my employment and that I shall not make any disclosure at any time thereafter without the prior, written permission of   [Company]   , except as I may be required to make such disclosure by judicial process or operation of law.

[OPTIONAL ADDITION AS TO PROPRIETARY INFORMATION OF OTHERS]

I further agree and affirm that I am not bringing or disclosing to   [Company]   , and will not in future bring or disclose, the secret or confidential information of others, including former employers, which I have agreed or am otherwise bound not to disclose.

Date:_____   _____
                                                        (Signature of employee)

_____   _____
(Signature of witness)   (Typed name of employee)

_____   _____
(Typed name of witness)   (Payroll, badge, or
                                                         Social Security number)

_____
(Payroll, badge, or
Social Security number)

Nonemployee Secrecy or Nondisclosure Agreement

AGREEMENT made this _____ day of _____, 19____ , between   [Name and legal identification of organization disclosing the information]   , hereinafter referred to as the Company [or other appropriate designator], and   [name and legal identification of the person or organization receiving the disclosure]   , hereinafter referred to as the Supplier [or other appropriate designator],

WITNESSETH,

WHEREAS, the Company is the owner of certain secret and confidential information, including Trade Secret information, which is of great value and importance to the Company, and

WHEREAS, the Company desires to obtain and the Supplier desires to render (or sell) the products (or services) of the Supplier, under additional terms and conditions recited elsewhere, and

WHEREAS, the Company may disclose to the Supplier certain of its secret and confidential information, including Trade Secret information, for the sole and exclusive purpose of permitting the Supplier to perform the services or deliver the products hereinbefore noted,

NOW THEREFORE, in consideration of the mutual promises and undertakings herein made it is mutually covenanted and agreed by the parties as follows:

1. Supplier agrees that all disclosures of secret or confidential information of the Company will be received for the sole and exclusive purpose of enabling Supplier to render the required services or deliver the required product;

2. Supplier agrees that all secret and confidential information disclosed to Supplier by the Company remains the property of the Company;

3. Supplier agrees not to disclose to third persons any secret or confidential information disclosed by the Company without the prior, written permission of the Company;

4. Supplier agrees not to disclose to any of its own employees except those who have a need to know in order for the Supplier to perform the terms of this Agreement or such other terms and conditions elsewhere recited regarding the subject of this agreement, any secret or confidential information of the Company;

5. Supplier agrees to inform those of its employees to whom Supplier must disclose secret or confidential information of the Company's of the nature of such information and of the Supplier's and his employees' duty to safeguard such information and not to disclose it except in strict accordance with the terms of this agreement;

6. Supplier agrees to protect and safeguard all secret and confidential information disclosed to it by the Company in accordance with the security rules of the Company, receipt of a copy of which is hereby acknowledged;

7. Supplier agrees upon demand by the Company, and in any event upon the termination or completion of the rendition of services or delivery of product referred to herein, that Supplier will return to the Company all secret or confidential information in documentary or other tangible form, including all copies or reproductions, in Supplier's possession or under Supplier's control and will certify that all such information has

been returned to the Company or has otherwise been disposed of in accordance with this agreement;

8. The Company agrees that all disclosures by it to Supplier of secret or confidential information will be appropriately identified and that all such information in documentary or other tangible form will be accompanied by duplicate receipts of which one will be executed and returned to the Company by the Supplier;

9. The Company agrees that the terms of this agreement will not apply to any secret or confidential information already known to the Supplier to be public knowledge or which during the performance of this Agreement or thereafter becomes public knowledge by some means other than disclosure by Supplier.

IN WITNESS WHEREOF the parties have hereto affixed their hands and seals as of the day and year first above written.

—————————————————————l.s.   —————————————————————l.s.
　　　　　The Company　　　　　　　　　　　　　The Supplier

by: ——————————————   by: ——————————————
　　　　　(Signature)　　　　　　　　　　　　　(Signature)

——————————————   ——————————————
　　　　　(Typed name)　　　　　　　　　　　　(Typed name)

——————————————   ——————————————
　　　　　(Title)　　　　　　　　　　　　　　(Title)

ATTEST:　　　　　　　　　　　　ATTEST:

　　[If a corporation]　　　　　　　　[If a corporation]

Employee Agreement Not to Compete with Former Employer

AGREEMENT made this_____ day of_____ 19 ____, between _____ [name and legal identification of employer] ____, hereinafter referred to as the Employer, and _____ [name and identification of the employee] ____, hereinafter referred to as the Employee,

WITNESSETH,

WHEREAS, the Employer employs or will employ the Employee under terms and conditions not all of which are stated herein, and

WHEREAS, the Employer owns and utilizes secret and confidential information, including Trade Secret information, developed at considerable expense and of great value to the Employer, and

WHEREAS, the Employer will disclose such secret and confidential information to the Employee, from time to time, for the sole purpose of enabling the Employee to perform the duties of his employment with the Employer, and

WHEREAS, the relationship between the Employer and the Employee is one whereby the Employer reposes special trust and confidence in the Employee,

NOW THEREFORE, in consideration of the employment or continued employment of Employee by Employer, and of the mutual promises herein made, it is agreed and covenanted between the parties as follows:

REGARDING NONCOMPETITIVE EMPLOYMENT

1. For the duration of Employee's employment with Employer, and for an additional period of two years next following the termination of such employment, the Employee shall not independently or in association with others engage in any work for, accept any employment with, or render any service or advice or assistance, directly or indirectly, to any person, partnership, corporation, association or other organization engaged in the _____ [here recite the industry, business or activity covered by this Agreement] ___; or, for the duration of the same period, accept employment with any other person, partnership, corporation, association or other organization, in any capacity involving the duties of _____ [here recite specific jobs or occupations Employee may not hold] _____ to be performed with respect to _____ [here recite the specific product(s) or item(s) with regard to which Employee may not accept any of the just-described jobs with any employer] _____ .

2. If the Employee, in observance of the immediately preceding Clause 1, and solely because of the obligation imposed by it, and after the continued exercise of due diligence, evidence of which has been submitted to the _____ [person or official] ___ of Employer, shall be unable to obtain employment consistent with his training, education and experience, then Employer, in its sole discretion, shall either waive the observance of this agreement or pay to the Employee for the portion of said two-year period following Employee's termination of employment with Employer during which Employee is unable to find suitable other employment compensation in such amount that when combined with such other income as the Employee may have from employment, the total income from employment will be equal to _____ [recite amount] _____ percent of the amount Employee had been receiving from Employer as his base pay at the time of the termination of employment, exclusive of any extra pay or compensation or employee benefits. Compensation payable hereunder shall be computed from the date of termination of Employee's employment with Employer or the date on which any severance compensation being paid to Employee by Employer ceases, whichever is the later.

REGARDING ASSIGNMENT OF THIS AGREEMENT

3. This Agreement supersedes all previous agreements between Employer and Employee concerning noncompetition. This Agreement shall not be assigned by Employer without the consent of Employee. [This provision would be appropriate in situations in which the employee might not be willing to execute the agreement with unknown future assignees of this employer.]

REGARDING CONTROVERSIES

4. Any controversy or claim arising out of or in relation to this Agreement or the breach thereof shall be settled by arbitration in accordance with the rules of the ____ [American Arbitration Association or other recognized arbitration authority] ___ , and judgment upon the award of the arbitrator(s) may be entered in any court having jurisdiction thereof. [This clause would permit arbitration rather than judicial trial of a breach of the agreement. Arbitration could be faster, less expensive, less formal, conducted by experts in the industry or activity, and highly secret. Frequently, although not always, arbitration is a preferable resource for resolving disputes.]

IN WITNESS WHEREOF the parties have set their hands and seals hereto as of the year and day above first written.

—————————————————l.s.   —————————————————l.s.
(Signature of employee)   (Signature of employer—if corporate, use customary form)

Patent Agreement and Assignment of Inventions

IN CONSIDERATION of my employment by __[Company]__ I, the undersigned, understand and agree that any and all compensation received by me by reason of such employment constitutes full consideration for my obligation to perform the duties of such employment and full consideration for this Agreement and for the ownership by __[Company]__ of all my right, title and interest to and in any and all inventions, discoveries, ideas, and improvements, of any sort, whether or not patentable in the United States or elsewhere which, during any period of employment with__[Company]__I have in the past conceived, developed or perfected, or may in the future conceive, perfect or develop, whether alone or jointly with another or others, and either during or outside the hours of such employment, and either with or without the use of property and materials of__[Company]__ , and either on or off the premises of__[Company]__ , and which pertain to any activities, business processes, equipment, materials, or products in which   [Company]   has any direct or indirect interest whatsoever. Inventions, discoveries, ideas, and improvements, if any, constituting exceptions to this Agreement and not covered by its terms are listed or described on the reverse hereof or addendum hereto. Absence of any exceptions on the reverse or of any addendum indicates that there are no exceptions to this Agreement.

I further agree as follows:

1. I hereby grant to   [Company]   , its successors and assigns all my right, title and interest in and to any and all inventions, discoveries, ideas and improvements, above recited, except those, if any, described on the reverse or an addendum hereto;

2. I agree to disclose without delay to   [Company]   and its successors and assigns any and all such inventions, discoveries, ideas and improvements;

3. I agree to execute, at the option and request of   [Company]   , its successors and assigns all applications or other papers for the securing of patents, trademarks, copyrights and any and all other forms of protection with regard to any and all inventions, discoveries, ideas and improvements made by me;

4. I agree to execute such further assignments and transfers of my entire interest in any and all such inventions, discoveries, ideas and improvements to   [Company]   , its successors or assigns, as it or any of them shall request, and to do such other act and things as shall be so requested to make this agreement effective with respect to any such invention, discoveries, ideas or improvements.

WHEREUNTO I have set my hand and seal this _____ day of _____ , 19____ .

_____
(Signature of employee)

_____
(Typed name of employee)

_____
(Payroll, badge, or
Social Security number)

WITNESS:

_____
(Signature of witness)

_____
(Typed name of witness)

_____
(Payroll, badge, or
Social Security number)

Summary of a Business Ethics Survey

An attempt was made in a survey by the *Harvard Business Review* to determine what was considered by executives to be ethical and unethical information collection methods. Twelve situations were presented. The responses are summarized from the report ("Problems in Review—Industrial Espionage," Nov.–Dec. 1959).

	Percentage Approval
A retailer sends someone to "shop" in a competitor's store to get product and pricing information	96
An oil company establishes a scout department to watch the drilling activities of competitors	71
A company, learning of a competitor's test market, quickly puts on a special sale in the same location	64
A key employee is hired away from a competitor	59
Sales manager wines and dines his competitive counterpart, pumping him for information	47
Company representative poses as a prospective customer to get information from a competitor	32
A vice-president hires a detective agency to watch the proving grounds of a competitor	16
Design engineer steals the plans of a competitor's new model	4
Company president instructs his aide to secretly record conversations in a competitor's office	4
Production manager rewards a competitor's employee for certain process information	3
A company plants confederates in a competitor's organization	2
District manager wiretaps the phone of his local competitor	1

The interpretations of data as outlined in the article are quoted below:
•   Comparison shopping—practiced for quite some time by retailers and manufacturers of consumer goods—is the most emphatically ap-

proved. The similar, but more recent, practice of scouting is approved by a somewhat smaller majority of executives. In certain respects, both these activities can be considered aboveboard, since they occur in full public view.

• The third most approved practice—moving into a test market—raises an interesting problem. A company can readily learn of a competitor's test, through trade commentary, press releases, and salesmen. The question is whether or not to engage in counterintelligence.

• The fourth situation the executives approve of—by a scant majority, in this case—is that of hiring employees from a competitor. This issue contains both the possibility that the man is hired for his abilities and the possibility that he is hired for the information he has.

• As for the one topic on which the executives are almost evenly divided, further analysis by industry indicates a majority of executives in the manufacturing of industrial and consumer products, engineering, and transportation approve of wining, dining, and pumping competitor counterparts, while the majority in all other industries are against it.

• 52 percent of executives in the manufacturing of industrial products and 44 percent of those in the manufacturing of consumer products state that their salesmen furnish "much" or "extensive" information about competitors.

• 69 percent of executives in the retail and wholesale trades and 66 percent of executives in transportation indicate that their salesmen furnish "some," "much," or "extensive" information about competitors.

• 47 percent of executives in consulting and business services and 39 percent of those in banking feel that their salesmen furnish "little" information or "none."

• 52 percent of executives in banking, 47 percent of those in transportation, and 40 percent of those in communications and engineering obtain "much" or "extensive" information from published sources.

• 25 percent of those in wholesale and retail trade state that they receive "much" or "extensive" information from company suppliers or customers.

• 40 percent of executives in banking indicate that they receive "much" or "extensive" information from personal contacts with executives of other firms.

• Over 60 percent in every industry feel that their companies receive "some," "much," or "extensive" information from such personal contacts by executives.

The following table taken from the article gives further information concerning the amount of information about competitors obtained by companies from various sources. Numbers are percents of all respondents.

Source of Information	Amount of Information Obtained					
	None	Little	Some	Much	Exten-sive	Total
Company salesmen	10.1	12.6	36.6	25.6	15.1	100
Published sources	5.6	17.8	40.6	20.6	15.4	100
Personal or professional contacts with competitors	4.1	14.9	50.2	22.8	8.0	100
Company suppliers	14.2	26.7	45.0	12.2	1.9	100
Advertising agencies, consultants (excluding formal marketing research)	37.5	35.1	21.6	3.9	1.9	100
Hiring employees of competitors	52.7	33.7	10.9	1.9	0.8	100
Undercover activities by company employees	84.2	12.9	2.2	0.3	0.4	100
Undercover activities for company by outside agency	87.7	9.2	2.4	0.3	0.4	100

INDEX

communications (cont.)
  see also bugging and wiretapping; telephone; telecommunications connections
"company secret" classification, 86
competition
  industrial espionage and, 14
  reverse engineering and, 65–66
  trade secrets and, 33
  see also trade secrets
computer
  crypto systems and, 114–115
  data inputs to, 99–101
  electromagnetic bugging of, 103–104
  espionage and, 96–115
  information storage media for, 105–109
  machine time accountability for, 111
  plain-text data in, 99–100
  precautions in operation of, 109–111
  remote-access vulnerability in, 111–114
  sensitive files for, 107
  tape and disk storage for, 105–109
  telecommunications connections to, 110
  unauthorized service bureau personnel and, 97–99, 102–103
  vulnerability of, 98–99, 102–103
computer area, admittance to, 104–105
computer service bureau, vulnerability through, 97–99, 102–103
Concorde supersonic transport, 55–56
confidential information
  classification of, 84–85
  federal civil law and, 32–33

remedies available to proprietor of, 30–32
  trade secrets and, 24–41
contractor and vendor employees
  disclosure to, 93
  industrial espionage by, 49–50
  trade secrets and, 52
Cost Accounting Standards Board, 21
Crest toothpaste case, 10–11, 20
criminal law, state, 38–40
crypto systems, computer security and, 114–115
customer relations, as valuable asset, 20–21

data
  in process, 103
  magnetically inked or coded, 101–102
  see also information
Defense Department, U.S., 21
Devine, Matthew, 69
disclosure
  regulating of, 92–94
  unauthorized, 95
disloyal employee, 43–45
  see also employee
displays and exhibits, control checks for, 126
document classification, information control and, 125–126
  see also classification
document control system, 127
document cover sheet, 157
documents, locked containers for, 127
Drambuie liqueur, secret of, 18
Dun's Review, 146
Du Pont de Nemours, E. I., & Company, 61–62

East Germany, industrial spies in, 55
education, for employees, 119–120

Index

trash and scrap, information gathered from, 62
trespass, in industrial espionage, 59–60
Tribuno vermouth, 18
TU-144 Soviet airliner, 56

Ulbricht, Walter, 55
undercover operator, in industrial espionage, 58–59

Vanderbilt, William H., 144
Van Doren, Charles, 145
vendor employees, as security threat, 48–50

vendors, secrecy agreements with, 136–137
verbal information, control of, 127
visitors' logs, as protection device, 121–122

Wall Street Journal, 61
wide-area telephone service (WATS), 113
wiretapping
bugging and, 68–81
federal laws on, 78–80
state laws on, 80–81

Zildjan cymbal, 18